Critical Thinking for Medical and Graduate Students

Critical Thinking for Medical and Graduate Students

Jonathan M. Berman, Troy Camarata, and Tony A. Slieman

The MIT Press
Cambridge, Massachusetts
London, England

The MIT Press
Massachusetts Institute of Technology
77 Massachusetts Avenue, Cambridge, MA 02139
mitpress.mit.edu

The MIT Press would like to thank the anonymous peer reviewers who provided comments on drafts of this book. The generous work of academic experts is essential for establishing the authority and quality of our publications. We acknowledge with gratitude the contributions of these otherwise uncredited readers.

This book was set in Stone Serif and Stone Sans by Westchester Publishing Services. Printed and bound in the United States of America.

Library of Congress Cataloging-in-Publication Data is available.

ISBN: 978-0-262-55320-9

10 9 8 7 6 5 4 3 2 1

EU Authorised Representative: Easy Access System Europe, Mustamäe tee 50, 10621 Tallinn, Estonia | Email: gpsr.requests@easproject.com

Dedicated to our students, and mentors, who keep us thinking, and to our families

Contents

Introduction

You are driven by a burning passion to know, create, heal, innovate, experiment, adapt, learn, and solve the problems of patients and the world. Passion is what fuels your productivity, knowledge, creativity, and persistence. Passion is what brought you to a career that challenges you intellectually, socially, and sometimes physically. You care deeply about what you do. That's why you're reading this book.

Your passion deserves to be supported and refined into a useful tool through the process of learning. As we discuss topics essential to critical thinking in this book, we will argue that critical thinking, a kind of slow, reflective, and effortful thinking, is integral to the learning process. Critical thinking will allow you to direct your passion into a tool to achieve your career and life goals.

The philosopher and educational reformer John Dewey saw critical thinking as a process for taking deepening consideration of a matter. Rather than accepting their first instinct or belief, a critical thinker pauses, considers options, tests ideas, and only then arrives at a conclusion. In this book, critical thinking is a kind of thinking that reflectively evaluates beliefs and options, tests hypotheses, and does not unreflectively choose the first option available.

Our goal won't be for you to engage in critical thinking at all times. Human beings will avoid thinking if possible. Thinking is an unreliable, slow, and expensive process, and if our brains can supplant it with a simpler mechanism or task, they will. Most of the time we rely on memory and preexisting schemata, effortlessly completing complex tasks like walking, reading, and knowing store hours without much conscious thought. Even small disruptions that require thought, like lost keys, forgotten passwords, navigating unfamiliar cities, or adjusting to new office layouts, can cause exhaustion.

Although we tend to avoid thinking, successful problem-solving and thinking, which are closely related to learning, can be very rewarding. To curious people, slow learning can be frustrating and effortful. Most of us enjoy solvable tasks like crosswords and karaoke over cryptographic puzzles or music composition. When the gap between what we know and what we need to learn is too high, we perceive low odds of success and we're less likely to try. However, with dedication and practice, we can master complex skills and be rewarded by the experience, like a student who commits to practicing music and eventually finds composition gratifying.

By teaching ourselves some of the science of learning, we can avoid some pitfalls that make the effortful process of learning seem less likely to succeed. We can learn to scaffold our own education, and turn the metaphorical cryptograms of our advanced education into easily solvable crossword puzzles. We can make unscalable mountains of thought into strollable molehills.

1 How Do We Know What's True?

> The greatest enemy of knowledge is not ignorance; it is the illusion of knowledge.
>
> —Stephen Hawking

The most important question in critical thinking is: How do you know something is true? Understanding *epistemology*, the study of *knowledge*, is critical for evaluation of evidence, for understanding which claims about science and medicine are well justified, and for conducting or interpreting scientific research without making unsupported or unsupportable claims.

What Is Knowledge?

One definition of knowledge (sometimes attributed to Plato [Dutant 2015]) is that knowledge is a *justified, true belief*. If something is *known*, a person believes it to be true; they have justification for that belief, and it is true. While indoors, I might believe that it is raining outside. I may have justification for this belief, such as consulting a weather report or seeing someone who just entered the office with a wet umbrella. However, if it had stopped raining twenty minutes ago, did I *know* that it was raining?

Usually, belief is used to denote acceptance of a concept, or confidence in its truth. So, to know something, we must have confidence that it is true, and it must be true.

A school of philosophy named *skepticism* has questioned whether it is possible to arrive at "knowledge" at all because it is not possible to eliminate all possible sources of doubt. You might trust your senses, but there are circumstances in which you might hallucinate. Some have even proposed that our reality might be a simulation like a computer game. Others have challenged the justified true belief idea of knowledge with counterexamples.

In 1963, philosopher Edmund Gettier published a paper that provided counterexamples to the justified true belief view of knowledge. In a "Gettier case," someone can hold a true belief based on faulty reasoning: You see someone out on the street wearing a red hat. You believe your friend Fatima is on the street because she always wears red hats. You run down the street, and you do see Fatima, but she isn't the one wearing the hat.

You had a justified true belief, but reasoning was faulty (Gettier 1963). Is this knowledge? Perhaps a justified true belief must not depend on false premises. We could add complexity to our definition of knowledge, or use different approaches.

In biomedical sciences, evidence-based and science-based medicine (EBM and SBM) are favored because they systematically test medical interventions using empirical evidence from scientific studies. This offers strong justification for belief in treatments due to reproducibility, peer review, and hierarchies of evidence.

How Scientists and Physicians Gather Knowledge

Scientific knowledge is different from *day-to-day knowledge*, and requires more rigor and higher standards of evidence. Science is the most reliable process for producing useful knowledge. Physicians and researchers acquire knowledge by many means, including patient history, experiments, data analysis, and literature review. Each provides different levels of justification, but are not equal in reliability.

Our goal is to eliminate poorly justified beliefs and retain well-justified ones. We need to set a high threshold for "knowledge" as this will impact diagnosis and treatment selection in medicine, and in all aspects of research.

How Scientific Fields Gather Knowledge

Scientific fields develop over time as knowledge changes and new information becomes available.

For example, the diversity of life on Earth developed over time through evolution by natural selection, genetic drift, and gene flow. How do we know that this is true? Our understanding of evolution developed over many decades, and went down wrong turns with neo-Lamarckism, orthogenesis, and saltation. Our understanding needed the discovery of Mendelian inheritance, the maturation of statistics, and the development of molecular biology to develop to its current state.

Science is conducted with different ways of acquiring and analyzing knowledge, and subject to constant revision and self-skepticism. Often, this is piecemeal, with many wrong turns, but self-corrects over time.

Normative Epistemology

Sometimes we learn that our beliefs were wrong. Deciding how to arrive at true beliefs is a job for *normative epistemology*. Science has proven itself to be one of the most effective means of arriving at true beliefs, but we must define it.

Most likely you have been taught about the scientific method and that it consists of linear steps. These steps are usually something like "identify a problem, make observations, develop a hypothesis, gather data to test that hypothesis, revise the hypothesis, and then repeat."

"The" Scientific Method

Scientific research is not always cleanly linear. Research can be messy, with failed experiments and wrong paths. Some fields like astronomy may not fit this model when discoveries sometimes precede hypotheses.

A hierarchy of observations, hypotheses, theories, and laws, such that one leads to the other (McComas 1996), always in that order, is not always the case. For example, Newton developed a law of gravity, but without proposing a hypothesis for its cause.

Sociological observations of scientists reveal a broader framework, involving creativity, prior knowledge, and perseverance. It has been suggested that "the scientific method" itself, propagated by science textbook, in the manner that Steven Jay Gould described erroneous propagation of the "Fox terrier comparison" (Gould 1991), where a phrase is repeated so often that it is retained in textbooks despite no longer being relevant, should not be taught, but rather that there are plural *scientific methods*. However, not all methods are scientific. Science is successful because it works, while other means of knowing such as divine revelation or guessing lack the same reliability and rigor.

Is Science "Special"?

There are many ways of learning about the world. Some methods are more effective than others at producing reliable, useful, and "true" results. We can call these methods "science," but what do these methods have in common that separates them from astrology? Why should a physician be granted more trust than a chiropractor?

Pursuing science as an endeavor and using science as a tool for the treatment of disease are premised on science being the best means of achieving an understanding of the world that makes reliable predictions, and in the case of medicine demonstrably improves outcomes for patients. Giving a word to those processes that are most effective toward this end (science) creates challenges. There are practitioners who want the prestige that scientific practices have earned, and therefore emulate some of the superficial accoutrements of science such as lab coats and scientific jargon. We need to understand what we should or should not call scientific so that effective practices can be more easily distinguished from ineffective practices.

Science and Its Boundaries: The Demarcation Problem

The question of what we can call science leads to further inquiry about *pseudoscience* (fake science) and activities that resemble science but lack its core principles. In medicine this question is crucial because fake treatments can harm patients and waste time and resources. For example, a prostate cancer patient taking chlorine dioxide, "miracle mineral solution," which is backed by anecdotes and not science, risks poisoning.

For working scientists, understanding what constitutes scientific inquiry avoids wasted effort and helps in the development of appropriate methods. While pseudoscience pretends to be science but lacks rigor or an evidence-based approach, nonscience can still have value, such as inquiry in the humanities, which is distinct from science.

Demarcation criteria (the criteria that let us say something is scientific) should be seen not as rigid conditions (Laudan 1983) but as a tool to identify modes of inquiry that are both fruitful and worthwhile. These distinctions can be seen as a continuum, with certain methods being more or less scientific. However, that does not mean that scientific and nonscientific practices are equivalent. A fuzzy border is a border. As Massimo Pigliucci suggests, "science" can be understood as a family resemblance between related practices, rather than a single entity (Pigliucci 2013). This view allows for a nuanced understanding of scientific inquiry and helps distinguish methods that work from those that don't.

Logical Positivism

Logical positivism was a philosophical movement that promoted the idea that scientific facts were those that could be verified either with sense experience, reason, or logical necessity. The division between truths verifiable by sense experience and those logical necessary is known as Hume's fork (Hume 2019). *Verificationism* was the demarcation criterion used by logical positivists who aimed to differentiate scientific inquiry from metaphysical speculation.

The hypothesis "If I let go of an apple, it will drop to the floor," can be "verified" by letting go of an apple. A logical positivist might reason from the specific to the general: "If I drop a pear, or a potato, they will both drop to the floor." With these further verifications, a general principle can be extrapolated: "Things that are let go of will fall to the floor." To a logical positivist, science is conducted by reasoning from past experiments and arriving at general principles. This seems to provide a demarcation between the activity of doing science and the activity of metaphysical speculation (guessing).

However, we only need a single instance of an object that does not fall to the floor, for example, a helium-filled balloon, to demonstrate that the principle does not hold true for all objects. "Any object I let go of will drop to the floor" is a false statement, even though we can verify it with an almost infinite list of non-helium-balloon objects.

If we wish to verify the statement "All swans are white," we might go to a pond and observe only white swans. We have "verified" the hypothesis. However, if we ever observe a black swan, the hypothesis was wrong. We could observe every swan in the world, save one, and the hypothesis could still be shown to be wrong. This is known as the *problem of induction* (Popper 1959).

A lot of scientific activity in the biomedical sciences looks superficially like verificationism. A paper might make a claim that such-and-such protein binds with so-and-so protein. Multiple figures and techniques might seem to verify this claim. However, none of these methods has *verified* this claim as true; they have only lent credibility to it. A single experiment of another type might show that the two proteins do not associate.

Falsification

Karl Popper proposed a solution to the problem of induction: We can never build scientific knowledge by making "verifiable" statements because a single observation could *falsify* them. However, many attempts to falsify a hypothesis, each of which fails, helps to *corroborate* the hypothesis.

Here, the demarcating problem is solved by a *falsifiability* criterion: "For an idea to be scientific, it must be possible to design a test that could potentially show that idea to be wrong." The goal of scientists becomes not to verify their hypotheses but to develop tests that could potentially show those hypotheses to be wrong.

A statement that might be non-falsifiable in one era (and thus not scientific) might become falsifiable in a later era when new tools become available to test the hypothesis. Likewise, fields of inquiry that are non-scientific, like alchemy, can over time transmute into scientific fields (chemistry) through the development of testable hypotheses and rejection of falsified ideas.

However, falsification has been criticized because it doesn't provide a means of selecting the *best* hypothesis. Falsification also isn't descriptive of how scientists have practiced science in the past. Often, theories can be modified to accommodate falsifications, if they have other things going for them.

Falsification is not claimed to be a means of identifying the best hypothesis, but rather a means of identifying whether a statement can be investigated with empirical science at all. Hypotheses that are not true may be falsifiable and therefore empirically scientific, and may even be corroborated by failed falsifications. Thus, no scientific hypothesis or theory can ever be said to be proven. Rather, certain ideas in science are very well corroborated. The principle of *fallibilism* holds that it is possible to hold beliefs as true with the acknowledgment that this truth is provisional.

Falsificationism alone isn't sufficient to determine which ideas should be believed. Failing to falsify a theory many times doesn't lend probability to it being true, yet we might structure our arguments around our failures to falsify an idea, and we might begin to believe theories that have failed falsification many times, with the provision that they might be falsified in the future.

Another challenge to falsification comes in the form of the Duhem–Quine problem. Simply put: In order to formulate a hypothesis, a number of logical assumptions need to be made, which are actually hypotheses themselves. For example, consider that observing a black swan doesn't necessarily falsify the "All swans are white" hypothesis, if we consider that we may have misidentified the bird, or our perception of color may be flawed. This problem tells us that we can't ever falsify a hypothesis in isolation from its underlying assumptions. In the early nineteenth century, the measurements of Uranus's orbit were not consistent with Newton's laws. This might have been taken to falsify Newton's laws; however, an alternative hypothesis emerged: the observation was of another yet-undetected planet, which is now known as Neptune. Another way of phrasing this is that science is provisional, and even falsifications are subject to later reinterpretation if an underlying assumption is challenged.

Many of the weaknesses in the scientific demarcation criteria are partially addressed through what has been called "consilience" or "triangulation." Hypotheses tend to gain credibility when they are supported by multiple, independent lines of evidence. When multiple kinds of evidence point to the same hypothesis, a shared flawed assumption or confounding variable will have fewer opportunities to create illusory effects.

If we were to consider the idea that "science cannot prove anything" so literally that we adopted *no beliefs*, and held *no assumptions*, then the practice of science would be impractical because we would be forced to spend all of our time hedging. Some claims are so well evidenced, have failed falsification so many times, and fit so well with multiple lines of evidence, that although we can acknowledge that a better idea might come along, it makes sense to accept those ideas as well justified, and believe them provisionally.

Falsificationism is a way to demark *empirical science* (Maxwell 1972). There are in fact multiple demarcation problems: the demarcation of empirical theories from metaphysical, the demarcation of science from non-science, the demarcation of non-science from pseudoscience, and the demarcation of worthwhile scientific inquiry from less-worthwhile scientific inquiry.

Some things that we generally consider to be science do not fall under the umbrella of empirical science. For example, Popper believed that statements about existence such as "White crows exist" are not falsifiable, and therefore not a part of empirical science. However, the discovery and description of a new species of crow are seen as a part of science. Falsificationism is useful for understanding how science goes from collections of individual observations to corroborating universal statements of theory such as "Natural selection is a major driver of the diversity of life on Earth," with attempted falsifications of predictions made by these theories.

A lasting criticism of strictly using falsifiability as our only means of demarcation is that it could allow certain pseudosciences to be called "scientific." Astrology makes claims about the world, and you could design a test that might demonstrate those claims to be false. If falsifiability is our only way of separating science, then pseudosciences like astrology might be included.

In practice, we want a means of demarcation that can exclude practices that continue to hold to claims even after they have already been falsified.

Scientific Revolutions

The Structure of Scientific Revolutions by Thomas Kuhn examined some of the history of chemistry and physics prior to 1850 and 1920, respectively, and proposed that science advances through what Kuhn called *paradigm shifts* (Kuhn and Hacking 2012). The term *paradigm* is vague (Masterman 1970), but in general paradigms are templates for understanding information.

An example is the geocentric model, which describes how the large bodies in the solar system travel around the Earth. Most astronomers worked under this belief for centuries, doing research that resolved discrepancies between data and the paradigm through increasingly complex models. Kuhn called the process of doing research to resolve these anomalies "normal science." Eventually, the anomalies became too large to ignore, and the heliocentric model where the large bodies in the solar system orbit the sun was (after initial resistance) adopted as a new paradigm.

Although *The Structure of Scientific Revolutions* is popular in science studies, it is not particularly useful to scientists. If paradigms do exist, there's little reason to think that they always exist as discrete entities. The history that *Structure* presents is a "great man" model of history in which great discoveries are made by individuals but most "normal science" is irrelevant, especially outside its paradigm. Most discoveries involve the collaboration of large teams advancing knowledge in small but important ways. It isn't always possible at the time research occurs to know which discoveries will be important and which won't be, so *Structure* doesn't give us useful norms for how to perform research, and does not suggest a solution to the demarcation problem.

Both Popper and Kuhn recognized that there are sometimes periods when experiments in a field are largely in agreement, and sometimes other periods of what Popper calls "crisis" in which experiments disagree with major theories. Popper identified strategies that might be used to preserve old theories, such as ad-hoc hypotheses (see chapter 7 on hypothesis development) to make falsifying evidence irrelevant, altering definitions in a theory, or even skepticism of the character or skill of the scientist who conducted the research. Popper believed that these tendencies could be avoided by adopting falsificationism, while Kuhn believed that this behavior was inherent to science.

Kuhn did not prove a strong basis for *why* scientists might come to favor one theory over another. Popper believed that prevalent scientific theories changed when a previous theory was falsified and that new theories rose to prominence when they had passed more rigorous testing than previous theories. This leaves open the possibility that future theories can modify or replace past theories, but still explains why these changes occur.

Kuhn observed that scientific advancement depends on the researchers, not just the state of the data. Good ideas might face challenges in being accepted until the right conditions exist for their acceptance. This does not tell us much about whether an idea should be accepted as scientific, but serves as a reminder that science is a *social endeavor* carried out by human beings.

Virtue Epistemology and the Scientific Attitude

Demarcation need not be a strict binary between "scientific" and "not scientific," but scientific efforts can be more or less scientific depending on properties that characterize those efforts. These properties might be termed "scientific virtues" (Paternotte and Ivanova 2017), and might include such things as maintaining clean glassware, taking detailed notes, attempting to falsify hypotheses, and doing statistical tests as appropriate to the limitations of those tests. The accumulation of scientific virtues makes an idea or process "more scientific," and the accumulation of vices such as logical fallacies or inaccurate data makes a process less scientific.

Attention can also be paid to what motivates scientists. Scientists expressing emotions like faith or certainty may be more likely to hold false beliefs than those who express emotions like curiosity, hope, and doubt.

Another approach is to consider science to be defined by a "scientific attitude." In practice, scientists don't always neatly follow all of the steps of the "scientific method," or of falsificationism, yet their process may still be considered scientific.

One articulation of this attitude was made by Richard Feynman in his 1974 Caltech commencement address (Feynman 1974): "It's a kind of scientific integrity, a principle of scientific thought that corresponds to a kind of utter honesty—a kind of leaning over backwards." Feynman was hoping to understand why some fields made significant practical progress, like physics, while other fields, such as education research, seemed to languish with at best modest improvements to outcomes. Those who did good science bent over backward to ensure that they were approaching their subject with utter honesty, controlling as many variables as they could, and taking nothing for granted. Fields that languished were doing what he called "cargo cult science," which imitated some of the accoutrements of science, but lacked radical honesty.

Lee McIntyre in *The Scientific Attitude* summed it up as two principles held by those with a scientific attitude: "We care about empirical evidence" and "We are willing to change our theories in light of new evidence" (McIntyre 2020). This can let in fields of inquiry that aren't traditionally considered to be science (like the rigorous and evidence-based study of history), but perhaps distinctions between activities like the evidence-based study of history and the evidence-based study of biology are more academic than practical. This makes the "scientific attitude" more useful for the demarcation of science from pseudoscience than science from non-science.

None of these criteria provide a strict line of demarcation between science, non-science, and pseudoscience. So, we can't logically prove that astrology is pseudoscientific. We can consider whether its practitioners care about empirical evidence, and whether they are willing to update their views. Purely psychological factors can seem unsatisfying as a demarcation criterion, but because science is conducted by humans, they may be the best we can do. Consideration of the attitude of practitioners is appealing because an examination of a field like homeopathy or astrology might show that their practitioners have continued to make claims that have been falsified, are unwilling to update their beliefs to new evidence, and therefore are pseudoscientific. A practice such as the discovery of new species easily passes this criterion because the people who perform this task care about evidence and update their theories when new evidence becomes available. Likewise, someone who adopts a view that the Holocaust never occurred, and refuses to update this view with new evidence, is doing a kind of pseudo-history, in the same way that people who don't care about medical evidence (or only pretend to care) might find themselves doing pseudoscience.

Conclusion

Understanding what demarks science from non-science may require an appreciation of both *logical* and *psychological* methods of demarcation. Psychological methods of demarcation present us with ideals of attitudes to strive for when conducting scientific inquiry, and analyzing and presenting data. Logical methods present us with methodological and experimental ideals to strive for when developing, conducting, and analyzing experiments.

Instead of treating science as a recipe in which steps are followed, science can be treated as a set of epistemological norms that emphasize the importance of evidence, critical communities, openness to changing minds, and, above all, striving for intellectual honesty and error correction.

Every experiment a scientist undertakes is done within a framework of testing hypotheses, answering a scientific question, or in some way producing knowledge. Likewise, science-based medicine requires the same commitment to truth and radical honesty, and a scientific attitude that is required of scientific research.

Summary

- Scientific knowledge is provisional and can change with new evidence or new reasoning.
- One of the most important questions to ask about any kind of knowledge claim is: "How do you know this?"
- Science requires a kind of radical honesty and interrogation of one's own methods, beliefs, and biases.
- Scientists are human beings who can be influenced by their culture and personal biases.

- Pseudoscience adopts trappings of science, but not its methods.

- It's difficult to know when there's enough evidence for a new theory to supplant an old theory.

- Scientific theories can't be "proven" or "verified," only corroborated or rejected.

Exercises

1. Do you believe that there is a difference between scientific knowledge and everyday knowledge? If so, what is that difference?

2. A list of different types of evidence follows. What order would you put them in, from most likely to influence your beliefs about a scientific question to least likely to influence your beliefs?

   Photographic evidence, PCR results (single trial), an audio recording, a single blood pressure measurement, multiple blood pressure measurements across a day, evidence accepted by all scientists in a field and backed by many peer-reviewed studies, evidence accepted by all scientists in a field but backed by few studies, a case report, divine revelation, something revealed in dream, a case report, something published in an open-access journal, a paper published in *Science*, an article published in *Scientific American*, information published in the *Scientific American* blog, statistical analysis, a single first-hand account, a double-blinded peer-reviewed trial, a single-blinded peer-reviewed trial, speculative explanation, common knowledge

3. In the following scenario, how do you respond, based on how you think the Food and Drug Administration (FDA) arrived at its beliefs about approved COVID-19 vaccines?

   a. You encounter a family member who refuses to be vaccinated against COVID-19, and states, "The vaccine was developed too quickly, I don't want to be a lab rat they're testing."

   b. How would your response change if this was a patient?

Works Cited

Dutant, J. 2015. "The Legend of the Justified True Belief Analysis." *Philosophical Perspectives* 29: 95–145.

Feynman, R. P. 1974. "Cargo Cult Science." *Engineering Science* 37: 10–13.

Gettier, Edmund L. 1963. "Is Justified True Belief Knowledge?" *Analysis* 23, no. 6: 121–123.

Gould, S. J. 1991. "The Case of the Creeping Fox Terrier Clone." In *Bully for Brontosaurus*, 155–167. W. W. Norton.

Hume, D. 2019. *A Treatise of Human Nature*. e-artnow.

Kuhn, T. S., and I. Hacking. 2012. *The Structure of Scientific Revolutions: 50th Anniversary Edition*. University of Chicago Press.

Laudan, L. 1983. "The Demise of the Demarcation Problem." In *Physics, Philosophy and Psychoanalysis: Essays in Honour of Adolf Grünbaum*, edited by R. S. Cohen and L. Laudan, 111–127. Springer.

Masterman, M. 1970. "The Nature of a Paradigm." In *Criticism and the Growth of Knowledge*, edited by Imre Lakatos and Alan Musgrave. Cambridge University Press.

Maxwell, N. 1972. "A Critique of Popper's Views on Scientific Method." *Philosophy of Science* 39: 131–152.

McComas, W. F. 1996. "Ten Myths of Science: Reexamining What We Think We Know About the Nature of Science." *School Science and Mathematics* 96: 10–16.

McIntyre, Lee. 2020. *The Scientific Attitude: Defending Science from Denial, Fraud, and Pseudoscience*. MIT Press.

Paternotte, C., and M. Ivanova. 2017. "Virtues and Vices in Scientific Practice." *Synthese* 194: 1787–1807.

Pigliucci, M. 2013. "The Demarcation Problem: A (Belated) Response to Laudan." In *Philosophy of Pseudoscience: Reconsidering the Demarcation Problem*, edited by M. Pigliucci and M. Boudry. University of Chicago Press..

Popper, K. R. 1959. *Karl Popper: The Logic of Scientific Discovery*. Springer.

2 The Social Dimension of Science

Science is not a body of facts. Science is a method for deciding whether what we choose to believe has a basis in the laws of nature or not.

—Marcia McNutt

While most of this book focuses on the individual thinker, there is a social dimension of science. This social dimension can be examined from several viewpoints: the social aspects of the scientific enterprise, scientific consensus building, and the social conditions needed for science to occur.

Some have even argued that science is a "social construct." Understanding the precise meaning of this phrasing is key to understanding why it is mistaken (or at least not very useful). Thinking about social constructs flows from the idea that certain "realities" are independent of humans and would exist without them, and other realities would not. For example, a $5 bill holding a value of $5 is not a fact of nature; it is dependent on humans agreeing to its value. On the other hand, regardless of how humans classify it, there is an object in the solar system that we called Pluto, and it would exist in nature regardless of what we think of it.

Arguments that science is "socially constructed" rest on a few points (Latour and Woolgar 1986). Research priorities are influenced by society and by the principles of funding sources. Researchers have cultural backgrounds and biases that influence the research they do. The practices undertaken by scientists, such as peer review and publication, are the results of social developments. The scientific consensus often develops in a social manner, such as through debate and correspondence.

None of these points is individually false. Scientists are in fact humans with their own biases and influences. It is also true that research priorities are often determined by short-term political shifts. In the United States, for example, funding agencies ultimately answer to elected officials.

However, some choose to interpret the idea of the "social construction of science" as meaning that science is not privileged above other ways of learning and that its outputs might vary from culture to culture. We have good reasons to believe that this interpretation is not true. For example, scientific predictions such as the dates of solar eclipses hold true across all cultures and can be independently tested; the laws of physics discovered through science hold true regardless of the culture of the investigator; and technological advances made possible through scientific discoveries (such as computers and smartphones) operate regardless of the beliefs of the user.

The Mainstream Scientific Viewpoint Is Sometimes Wrong

Just as scientists build consensus viewpoints and, in so doing, minimize individual error, mainstream scientific viewpoints have been wrong. In some cases this has led to persecution or marginalization of people whose unpopular scientific ideas later proved to be correct. For example, when the first paper describing *Heliobacter pylori* as the cause of peptic ulcers was first submitted, it was rejected and rated in the bottom 10 percent of papers seen by those reviewers that year. A Nobel Prize was later won for this work, but only after an author drank some of the same bacteria, causing himself to develop a peptic ulcer (Marshall and Warren 1984). In the long term, scientific errors tend to be corrected; however, even widely accepted beliefs and truths are subject to challenge by new data.

Failures of science to quickly eliminate false beliefs is also frequently tied to the social beliefs and who is conducting research. In the nineteenth and early twentieth centuries there were various attempts made by white scientists to find differences between human races. Now scientists understand that races are primarily social categories, that there is more within-group than between-group genetic variation, and that the easily visible traits commonly associated with race are poor markers for genetic diversity. Nevertheless, Samuel George Morton attempted to categorize races by skull size (and thus intelligence). Franz Joseph Gall helped to develop phrenology, the pseudoscientific use of skull shape to determine personality. Samuel Cartwright invented "drapetomania," a "mental illness" that caused enslaved people to want freedom. US-based eugenics programs and sterilization laws were imitated in more extreme forms by Nazi scientists, citing the "science" of eugenicists. Likewise, in the Soviet Union, the beliefs of one scientist, Trophim Lysenko, that plants of the same species could not compete with each other was advanced because it was seen as more politically aligned with communism than the beliefs of evolution by natural selection, in which capitalist-like competition caused some plants to die and others to thrive. The implementation of Lysenko's ideas resulted in crop failures that doubtless were a contributor to famines that killed millions.

Societies of Scientists

During the early development of science, scientists encouraged others to replicate their experiments. Results were shared publicly and debated in scientific societies. In many ways the advancement of scientific knowledge can be thought of as the advancement of the debate among scientists.

The "great man" theory is a way of viewing history as the outcome of "great men," or heroes whose individual actions led to the important events of history. (We acknowledge that the term used is sexist.) This is contrasted with the idea that historic events are the result of broader social and political factors. Individuals who influence historic events are products of the societies they live in, and in another social context may have lived unremarkable lives.

Although histories of science often focus on individual scientists like Oppenheimer, Einstein, or Darwin, these are people who operated within specific scientific contexts. In many cases, scientific discoveries are made simultaneously by multiple research groups. Darwin published his theory of evolution by natural selection at the same time that Alfred Russel Wallace was working on his theory. Challenging this "great scientist" theory of discovery is the fact that science tends to advance in such a way that discoveries are made when the underlying scientific knowledge to make those discoveries has sufficiently advanced, hence the simultaneous discovery of calculus by Newton and Leibniz, since mathematics had at that point advanced in the ways necessary for calculus to be developed.

A more nuanced understanding of scientific discovery includes elements of both these viewpoints. On occasion, individual scientists make advances, but these advances are tied to the knowledge and work of other scientists, the giants on whose shoulders those scientists stand.

One of the markers that distinguish pseudosciences is that they frequently champion the beliefs of a single founder or iconoclastic thinker. Examples include Samuel Hahnemann, the founder of homeopathy; L. Ron Hubbard, the founder of scientology; Rudolf Steiner, the founder of anthroposophy; D. D. Palmer, the founder of chiropractic; and other late nineteenth-century founders of pseudo-medical disciplines. While some of the ideas championed by these founders have found mainstream or scientific acceptance, the majority are not based in science and lack mainstream acceptance. Despite this, the teachings that have failed scientific testing are still taught. In contrast, the social enterprise of science recognizes the fallibility of individual scientists. Even scientists thought to be great individual thinkers have made mistakes, for example, Einstein's rejection of quantum mechanics. Linus Pauling believed that high doses of vitamin C could cure many diseases, including cancer. While some scientists are still venerated despite some false beliefs, the culture of science allows for this without adopting the false beliefs.

The Social Factors That Must Exist for Science to Thrive

Another social dimension of science is the requirement for certain societal conditions to advance. These include the free exchange of ideas, the ability for other scientists to give and accept criticism, the ability to replicate experiments, the ability to make mistakes without punishment, robust education, and academic freedom and protection from censorship (Merton 1973).

Free Exchange of Ideas

The self-correcting and error-checking mechanisms in science rely on peer review to help identify errors, biases, and limitations of research. The process of scientific consensus-building relies on replication of others' results and the accumulation of knowledge, which cannot occur if the free exchange of ideas is hampered.

Additionally, the free exchange of ideas allows for scientists to work on the same problem at the same time, and can result in more rapid advancements. Additionally, it has been observed by many scientists that the intersections between fields are often fruitful areas for new discoveries.

Many of the challenges being addressed by scientific research such as climate change, global pandemics, and space exploration are globally relevant. Therefore, global contributions to scientific research are needed.

Giving and Accepting Criticism, Making Mistakes

For all of the reasons listed above, environments in which scientists are free to give and accept criticism are environments where science thrives. The culture of scientists often includes criticism that to outside observers can appear unduly harsh or direct, but this is a necessary factor for scientific advancement.

Failure is an expected outcome in scientific research. Not all methods or experiments will result in the intended outcome. In environments in which mistakes are not tolerated, not only can error correction not occur when those mistakes do inevitably happen, but additional incentives are created to falsify data.

Education and Academic Freedom

If scientists did not train future scientists, then the advancement of science would last only a single generation. The education of young scientists has become one of the major mechanisms of progress in professional science, where the scientific workforce is often composed of trainees, and much scientific research is conducted at universities.

Because many scientists are also academics, the conditions of the academic work environment have a significant influence on the scientific workforce. Mandates to align academic speech with particular political viewpoints is antithetical to the scientific need for absolute honesty (AAUP 1940). For roughly the past century, the strongest protection of academic freedom, allowing scientists to speak honestly without fear of losing their jobs, has been tenure. However, in recent years there have been political attacks on the tenure system, as well as changes in academic administration, which has meant more shifts to contingent employment, resulting in a chilling effect on scientific discourse.

Exercise

Imagine that a science popularization project you are working on gains unexpected attention, and a reporter for the *New York Times* contacts you to provide an interview. The reporter expresses concerns that a particular political party may feel challenged by the project. You respond, "Science isn't political," and explain further, but this is the only quote that's run. Internet users point out that the consequences of science can have political implications, citing examples of nineteenth-century race science, eugenics, and the Tuskegee experiment. How can you better express your ideas about the relationship between science and politics?

Works Cited

AAUP. 1940. "Statement of Principles on Academic Freedom and Tenure." *AAUP Bulletin*.

Latour, B., and S. Woolgar. 1986. *Laboratory Life: The Construction of Scientific Facts*. Princeton University Press.

Marshall, B. W., and J. R. Warren 1984. "Unidentified Curved Bacilli in the Stomach of Patients with Gastritis and Peptic Ulceration." *Lancet* 8390 (June): 1311–1315.

Merton, R. K. 1973. *The Sociology of Science: Theoretical and Empirical Investigation*. University of Chicago Press.

3   Bias: Lies You Tell Yourself

We don't see things as they are, we see them as we are.
—Anaïs Nin, *Seduction of the Minotaur*

The first principle is that you must not fool yourself and you are the easiest person to fool.
—Richard Feynman

We often see ourselves as unbiased, believing that if we had biases we would be aware of them and overcome them. However, we all have degrees of bias, and recognizing and combating biases is crucial for critical thinking.

A bias is a pattern in our thinking that is not rational. While the use of rationality is ideal, humans are prone to mistakes, which can be mitigated by intentional thought and identification of errors, although this rarely occurs naturally.

Cognitive biases are modes of thinking that don't align with reason. Psychologist Daniel Kahneman identified two thinking systems: *System 1* is fast, unconscious, and error-prone, and *system 2* is slower more thoughtful and effortful (Kahneman 2011). System 1 determines our immediate reaction to stimuli, for example, how to drive a car on an empty stretch of highway, or how to complete other often repeated tasks, like diagnosing internal bleeding in an emergency department (for an experienced physician). System 2 requires more thought, attention, and effort, such as attempting complex math problems, identifying errors, recognizing sounds, completing tasks that are rarely done. Many cognitive errors arise from relying on system 1 thinking when system 2 thinking is needed. For example, the *anchoring effect*: Imagine you are shopping at a store for a pair of jeans. The price on one pair is labeled "$20." Another pair is labeled "~~$40~~ $20." Most people will view the second pair as more valuable than the first, although the prices are identical. System 1 thinking has set a value to the jeans ($40) and a price for the jeans ($20). System 2 thinking would tell us that prices for the jeans are identical ($20 and $20), and therefore both pairs have the same value. Although placing more value on the "*anchored*" jeans is not rational, anchoring is so effective that it is used by almost all retailers to manipulate consumers.

Another framework to view cognitive biases is through *cognitive load theory*, and the development of *schemata*. In this model there are two kinds of memory, *working memory* and *long-term memory*. Working memory is limited in that it can only hold a limited number of "items," such as names, colors, faces, or numbers. Long-term memory is (nearly) unlimited. At any time the working memory can call forth items from long-term memory, but the number of items it can hold is still limited. Say, for example, you want to categorize trees in a forest. For each plant you could make a list of tree-like qualities (e.g., branches, leaves or needles, height, bark type) and go through the list to determine if each plant is a tree, and what type of tree it is. This would be an exhausting process, and our short-term memory would quickly reach its limits.

Instead, we develop *schemata*, shortcuts that can be stored in short-term memory that stand in for more complicated sets of ideas. Instead of having to exhaust all the spaces in short-term memory with "pine, oak, hickory, beech, maple," we can compress these into a schema "tree," which encompasses all "tree-like" qualities. When encountering a new plant, we need to only compare it to the tree schema to determine if it is a tree, and not all other plants previously designated as trees. Many plants with very different evolutionary histories such as gymnosperm trees and angiosperm trees can all be called "trees."

A chess master, having encountered many chess games, forms a schema for each common opening, and is able to quickly identify a next move without the exhausting work of exchanging much information with long-term memory. A novice chess player, lacking this schema, must think through each possible move to follow the opening. When both novice and master are presented with an opening neither has seen before, they both must think through moves, and take a similar amount of time to do so.

Cognitive load theory is used as a framework for understanding how people learn, and how to scaffold information so that learners can most effectively develop schemata in order to understand more complex ideas. (This will be expanded on in chapter 17.) However, it can also help us to understand errors in thinking and cognitive biases.

Forer Effect

The *Forer effect* causes people to rate descriptions of themselves as highly accurate when in fact the descriptions are very general and might apply to anyone (Forer 1949). This effect can serve as a partial explanation for how pseudoscientific ideas like astrology or the Myers-Briggs Type Indicator assessment gain widespread belief and acceptance. They purport to supply information that is specific to an individual, and because the information given is vague, they are able to fit that information into their own lives.

This effect is also used by fake mediums, psychics, and "intuitives" who claim to provide specific information. A psychic might make vague statements that apply to almost anyone—for example, "Somewhere in your home there is a box with loose photos"—and customers will leave with the impression that highly specific insights were provided.

Bertram Forer demonstrated this effect in 1949 by having his students take a personality test, after which each was given an identical assessment. The assessment contained statements like, "You have a great need for other people to like and admire you," "You have a tendency to be critical of yourself," "You have a great deal of unused capacity which you have not turned to your advantage." Consistently, students rated these statements as highly reflective of their personalities. Similar experiments have demonstrated this effect many times.

Confirmation bias describes a tendency to focus on evidence that confirms what we already believe, while ignoring contradictory evidence. The first studies of confirmation bias occurred in the 1960s. In science we typically formulate hypotheses, then design tests that disconfirm those hypotheses. If a hypothesis survives long enough, we may think of it as corroborated and grant provisional assent to that hypothesis. Confirmation bias describes the tendency to do the opposite of this, formulating a belief and then designing tests that can only reinforce our beliefs.

In an experiment, participants were tasked to supply a triad of numbers and to discover a rule on which the triad was based. "2, 4, 6" is an example triad. Their directive was that "the three numbers be presented in increasing order." The participants would write down a triad and the researcher would inform them if the triad conformed to the rule; eventually, the participants would have the option to tell the researcher that they believed they had found the rule. Participants would suggest triads like "8, 10, 12" and "36, 38, 40"; discover that both conformed; and announce that the rule was "numbers increasing by 2."

Frequently, participants would announce rules having only seen that the ideas they had tested did conform, but not having devised any triads that they knew did not conform to their hypothesis. If the participant believed that "numbers increasing by 2" was the rule and wanted to test that hypothesis, they could have suggested the triad "8, 9, 11." This could have disconfirmed their hypothesis. That is to say, participants tended to run tests aligned with their hypotheses, but not alternative hypotheses.

Confirmation bias manifests in other ways. Rather than designing tests that seem to confirm a hypothesis, a voter might seek out information sources that provide news favorable to their viewpoint. A social media user might block news sources or users that present alternative viewpoints. Confirmation bias occurs whenever we fail to think of reasons why our beliefs might be wrong. A variety of explanations for confirmation bias have been offered. One is that the reliance on heuristics and "type 1" thinking makes it costly or difficult to consider more than one hypothesis (Kunda 1999). Another is that there is a tendency to focus on positive ideas and to ignore negative ones. Still another is that rather than looking for information that will make us right, people are averse to being wrong, and avoid thinking in ways that will force them to change their viewpoints (Haidt 2012).

Confirmation biases are especially relevant to medicine because confirmation bias can reinforce ineffective treatments or incorrect diagnoses (Nickerson 1998). For centuries the science of medicine stagnated, with treatments that were not effective, yet working physicians employed them every day throughout their careers. Galen's medical texts, which ironically advocated for empiricism, were often followed without experimentation. Confirmation bias led to a focus on successful recoveries rather than failed ones, allowing flawed treatments to persist (see chapter 11).

Physicians often overestimate the likelihood of their diagnoses and struggle to revise probabilities based on new information. In science, confirmation bias can lead to "hypothesis myopia" (Nuzzo 2015), where we focus on evidence supporting a single hypothesis, while ignoring or discounting alternatives.

"Disconfirmation bias," or *asymmetric attention to detail*, occurs when a researcher engages in deeper scrutiny of information that seems to contradict a preferred hypothesis than information that seems to support it. Scientists will often discount disconfirmatory evidence as not "real," until it has been repeatedly demonstrated (Fugelsang et al. 2004). Statistical errors in scientific literature err toward supporting the researcher's hypothesis, likely due to this bias (Bakker and Wicherts 2011). Kuhn's argument in *The Structure of Scientific Revolutions* (see chapter 1) suggests that anomalies in the current "paradigm" must accumulate until it is impossible to discount them.

Hindsight bias causes us to see past events as more predictable than they actually were (Roese and Vohs 2012). A cardiologist may miss signs of a latent heart condition on an EKG. Later, after the patient has a heart attack, another cardiologist may examine the EKG and believe that the signs should have been obvious. The second cardiologist may be falling prey to hindsight bias, seeing the diagnosis as more straightforward than it was. Mechanisms for hindsight bias include *foreseeability, inevitability,* and *memory distortion.* Foreseeability describes the belief about one's own abilities, such as "I would have caught that on the EKG." Inevitability describes beliefs about the world, like "Of course she had an MI; nothing else could have happened." Memory distortion involves misremembering one's earlier predictions. For example, a third cardiologist may misremember her encounter with the patient—"Yes, I said she had that heart condition at the time. I don't know why it wasn't charted."

Availability bias affects probability estimation when easily accessible information seems most likely. During a period of high flu and low COVID-19, a physician who has been hearing a great deal of news about COVID-19 recently may be more likely to diagnose it over flu. In a 1992 study, physicians were shown case descriptions of patients with various symptoms and sexual preferences. Physicians were more likely to believe a patient had a more publicized disease (AIDS) than a more common disease (influenza) if the patient was identified as homosexual (Triplet 1992). Other studies have shown that physicians given information about dengue fever are likely to overdiagnose it shortly after. Likewise, research has shown that second-year residents are prone to availability bias in diagnosis (Li et al. 2020; Mamede et al. 2010). A common example is the fear of flying exacerbated by overestimation of the danger of air travel due to the ease of recalling accidents that have received media attention.

Post-hoc rationalization involves reaching a conclusion, then developing reasons later on to justify it. This often occurs in daily life when people make decisions based on nonrational processes like emotions. For example, the social psychologist Jonathan Haidt demonstrated this in experiments involving bizarre ethical scenarios. Participants would judge a morally ambiguous, but emotionally charged action as wrong, but would struggle when asked to explain why, often inventing reasons after the fact (Haidt 2001).

Motivated reasoning, a broader form of post-hoc rationalization, occurs when we make choices and then look for reasons to support those choices after. When students were told that they had tested negative or positive for a fake risk factor or a disease, those who were told they were at risk had a higher estimate of the false-positive rate of the test (Ditto et al. 1988).

Memory Biases

Memory biases are cognitive biases that alter or distort the recall of memory. Memory is not like a computer file, which will replay identically each time. Human memory is flawed and prone to distortion and biases.

Memory biases may be influenced by mental health. For example, those with social anxiety disorder may have enhanced memory for socially threatening situations (Mathews and MacLeod 2005). People with major depressive disorder often exhibit an explicit memory bias, remembering events with a negative emotional valence (things that make them feel bad) more readily (Everaert et al. 2022). Processes in our cognition can thus affect our recall.

Psychologist Daniel Schacter identified "seven sins" of memory (Schacter 1999):

1. *Transience:* Memories interfere with one another, with newer events being easier to recall (Schmolck et al. 2000).

2. *Absent-mindedness:* Forgetting things due to inadequate attention being paid during encoding.

3. *Blocking:* Previous memories interfere with developing new memories (Luque et al. 2018).

4. *Misattribution:* The tendency to remember information correctly but to misremember the source of the information.

5. *Cryptomnesia:* An idea may unconsciously influence development of an idea, but is viewed as original. There are famous examples of cryptomnesia such as George Harrison (the Beatle) inadvertently writing "My Sweet Lord" with a melody very similar to "He's So Fine." Helen Keller accidentally wrote a story very similar to a preexisting story.

6. *Source confusion:* The source of a memory of an actual event is mistaken. You might recall a story differently over time as it is recounted and remember the last time you told the story rather than the event itself.

7. *False memory:* Memory of events that did not happen.

The unreliability of memory can lead to the expression of other cognitive biases as well. Schemata concerning social class can influence memory, as in an experiment where participants recalled higher IQ scores for a girl pictured in front of a wealthier home (Darley and Gross 1983). *Salient objects*, which are surprising, noticeable, colorful, or otherwise remarkable, can draw attention and interfere with the memory. For example, when shown a video of someone approaching a bank teller with a gun or with a checkbook, those shown the gun are less able to identify the criminal later on (Loftus et al. 1987).

Scientific progress depends on not only caring about evidence, but updating beliefs when new evidence comes along. Although our natural inclinations may not favor such updating, trained scientists can improve their ability to change their beliefs in light of new evidence, though not as much as necessary

(McDiarmid et al. 2021). Just as we use tools such as recording devices to enhance our memories, which we know to be fallible, using tools to improve our thinking can help us enhance our ability to overcome cognitive biases.

Implicit and Explicit Bias

Implicit and *explicit biases* are "attitudes, behaviors, and actions that are prejudiced in favor of or against one person or group compared to another" (NIH n.d.). Implicit biases are unconscious associations between qualities and certain groups, such as race or gender. Explicit biases are conscious prejudices.

Many studies have been conducted using *implicit association tests*, which track factors such as reaction time to certain stimuli and accuracy with pairing terms to identify potential biases. These implicit biases may not be conscious on the part of the test-taker, but might influence how they interact with people and what decisions they make. For example, a test-taker might show an implicit unconscious association between masculinity and science, which could potentially impact hiring or grading decisions. The aggregate effect of such biases across many people and many such decisions could have significant negative impacts on people in these classes.

Such biases can be thought of as unreflective, or system 1, thinking. The scientist who scores as moderately biased toward associating science with masculinity might tell you that "of course women are as capable and accomplished as men in STEM fields," when they are able to engage system 2 thinking. However, simply making this person aware of their implicit bias may not be an effective debiasing strategy in its own right.

Debiasing

Debiasing is the process of reducing cognitive and implicit biases. Results on the use of training to reduce bias have been mixed; however, in most cases it is possible to reduce biases in specific processes and tasks. A poker player can be taught to make statistical rather than gut instinct judgments in card games, but these individual reductions in bias do not translate to other knowledge domains. However, training those in fields that have an emphasis on statistics and logic can have enduring effects (Nisbett et al. 1987).

In 1994 Wilson and Brekke proposed that debiasing can occur, but only when certain conditions are met. The person must be aware of their bias, they must be motivated to correct the bias, they must be aware of the direction and magnitude of the bias, and they must have the self-control to adjust their response (Wilson and Brekke 1994). Which of these prerequisites can be trained? This book is premised on the idea that improving awareness of biases can set up the conditions for a motivate, and disciplined person to overcome their cognitive biases.

Another approach to debiasing is to anticipate future biases and then create environments that encourage rational choices through "nudges." These nudges make rational choices easier, more appealing, or more obvious to make. Examples of nudges include clear labeling of food calories, placing your phone across the room to encourage waking up all the way, and building breaks into projects or using checklists to promote slow, system 2 thinking.

Conclusion

We are all affected by biases. Awareness of these biases is the first step toward addressing them and designing conditions that prevent us from acting in accord with those biases instead of reason.

Summary

- Cognitive biases are modes of thinking in humans that can make them prone to deviate from reason.
- Two important biases are confirmation bias and motivated reasoning.
- Confirmation bias means that people are apt to favor evidence that confirms previously held beliefs.
- Motivated reasoning occurs when we make decisions, then look for reasons to support those decisions.
- These biases can be overcome with scientific training.

Exercises

1. *Alternate Perspectives:* Often the same events can be interpreted in different ways depending on the person doing the interpretation. For example, consider the following situation:

   A patient with congestive heart failure and kidney disease is being treated in a hospital. The patient's cardiologist prescribes furosemide to reduce the fluid retention by this patient, since it is a very effective diuretic. The patient's nephrologist tells the patient not to take the furosemide because furosemide can worsen the patient's kidney disease. Both physicians have legitimate reasons to select the course of treatment that they do, informed by their perspective on the case.

   a. Is one physician "more right" than the other?

   b. How can the dispute be resolved?

2. *Identifying Bias:* In the following examples, identify the bias being employed. Why is it a kind of bias? How does it deviate from rationality? If you don't know the name of the bias, simply describe the deviation from rationality.

   a. Sue is tasked to identify the pattern in triplets of numbers, and given the opportunity to receive feedback on test triplets, she selects whether they fit the pattern or not. She is presented with the triplet 9, 11, 13. She tests the triplets 15, 17, 19, and 3, 5, 7. Both fit the pattern. She declares that the pattern is "consecutive odd numbers."

   b. Jesse decides to research whether chiropractic can cure his migraine headaches. He researches this by reading papers recommended by the website of a local chiropractor, and by watching TikTok videos.

   c. Two local newspapers have coverage of the same crime. In the first, a person is referred to as "murdered," and the "killer" was arrested. In the second, the words "killed" and "suspect" are used.

   d. Jamal has concerns about vaccinating his young daughter. Most sources he finds from the government, scientists, and physicians are pro-vaccination, but he decides to search out anti-vaccination sources for "balance."

3. *Implicit Bias:* Take the Harvard Implicit Association Test (at https://implicit.harvard.edu/implicit/takeatest.html), and select the Gender-Science IAT. Follow the directions on screen and complete the test. The test will tell you if your responses indicate a bias toward associating men or women with science and with liberal arts.

   a. Consider the outcome of the test. What does it say about your own implicit biases?

   b. Take some of the other IAT tests on this website. What do the tests say about your own implicit biases?

Works Cited

Bakker, M., and J. M. Wicherts. 2011. "The (Mis)reporting of Statistical Results in Psychology Journals." *Behavior Research Methods* 43: 666–678.

Darley, J. M., and P. H. Gross. 1983. "A Hypothesis-Confirming Bias in Labeling Effects." *Journal of Personality and Social Psychology* 44: 20–33.

Ditto, P. H., J. B. Jemmott III, and J. M. Darley. 1988. "Appraising the Threat of Illness: A Mental Representational Approach." *Health Psychology* 7: 183–201.

Everaert, J., J. N. Vrijsen, R. Martin-Willett, L. van de Kraats, and J. Joormann. 2022. "A Meta-Analytic Review of the Relationship Between Explicit Memory Bias and Depression: Depression Features an Explicit Memory Bias That Persists Beyond a Depressive Episode." *Psychological Bulletin* 148: 435–463.

Forer, B. R. 1949. "The Fallacy of Personal Validation: A Classroom Demonstration of Gullibility." *Journal of Abnormal and Social Psychology* 44: 118–123.

Fugelsang, J. A., C. B. Stein, A. E. Green, and K. N. Dunbar. 2004. "Theory and Data Interactions of the Scientific Mind: Evidence from the Molecular and the Cognitive Laboratory." *Canadian Journal of Experimental Psychology* 58: 86–95.

Haidt, J. 2001. "The Emotional Dog and Its Rational Tail: A Social Intuitionist Approach to Moral Judgment." *Psychological Review* 108: 814–834.

Haidt, J. 2012. *The Righteous Mind: Why Good People Are Divided by Politics and Religion*. Knopf Doubleday.

Kahneman, D. 2011. *Thinking, Fast and Slow*. Macmillan.

Kunda, Z. 1999. *Social Cognition: Making Sense of People*. MIT Press.

Li, P., Z. Y. Cheng, and G. L. Liu. 2020. "Availability Bias Causes Misdiagnoses by Physicians: Direct Evidence from a Randomized Controlled Trial." *Internal Medicine* 59: 3141–3146.

Loftus, E. F., G. R. Loftus, and J. Messo. 1987. "Some Facts About 'Weapon Focus.'" *Law and Human Behavior* 11: 55–62.

Luque, D., M. A. Vadillo, M. J. Gutiérrez-Cobo, and M. E. Le Pelley. 2018. "The Blocking Effect in Associative Learning Involves Learned Biases in Rapid Attentional Capture." *Quarterly Journal of Experimental Psychology* 71: 522–544.

Mamede, S., T. van Gog, K. van den Berge, et al. 2010. "Effect of Availability Bias and Reflective Reasoning on Diagnostic Accuracy Among Internal Medicine Residents." *JAMA* 304: 1198–1203.

Mathews, A., and C. MacLeod. 2005. "Cognitive Vulnerability to Emotional Disorders." *Annual Review of Clinical Psychology* 1: 167–195.

McDiarmid, A. D., et al. 2021. "Psychologists Update Their Beliefs About Effect Sizes After Replication Studies." *Nature Human Behavior* 5: 1663–1673.

National Institutes of Health (NIH). n.d. "Implicit Bias." https://diversity.nih.gov/sociocultural-factors/implicit-bias.

Nickerson, R. S. 1998. "Confirmation Bias: A Ubiquitous Phenomenon in Many Guises." *Review of General Psychology* 2: 175–220.

Nisbett, R. E., G. T. Fong, D. R. Lehman, and P. W. Cheng. 1987. "Teaching Reasoning." *Science* 238: 625–631.

Nuzzo, R. 2015. "How Scientists Fool Themselves—And How They Can Stop." *Nature* 526: 182–185.

Roese, N. J., and K. D. Vohs. 2012. "Hindsight Bias." *Perspectives on Psychological Science* 7: 411–426.

Schacter, D. L. 1999. "The Seven Sins of Memory: Insights from Psychology and Cognitive Neuroscience." *American Psychologist* 54: 182–203.

Schmolck, H., E. A. Buffalo, and L. R. Squire. 2000. "Memory Distortions Develop over Time: Recollections of the O. J. Simpson Trial Verdict After 15 and 32 Months." *Psychological Science* 11: 39–45.

Triplet, R. G. 1992. "Discriminatory Biases in the Perception of Illness: The Application of Availability and Representativeness Heuristics to the AIDS Crisis." *Basic and Applied Social Psychology* 13: 303–322.

Wilson, T. D., and N. Brekke. 1994. "Mental Contamination and Mental Correction: Unwanted Influences on Judgments and Evaluations." *Psychological Bulletin* 116: 117–142.

4   Unraveling Reason: Logic, Logical Fallacies, and Rhetorical Fallacies

Fallacies do not cease to be fallacies because they become fashions.
—G. K. Chesterton

Logic

Popular culture often pits logic against emotion. In *Star Trek*, Spock wrestles between his emotions and his philosophy of "logic." However, logic is not itself a philosophy of life, or the opposite of emotion. It is a set of mathematical tools for understanding arguments and determining if the conclusions of an argument must follow from the premises. Logic gives us the ability to determine if our reasoning makes sense or not. Logic isn't sufficient on its own to give correct answers; an argument can be logically valid but based on false premises. However, combined with an analysis of the premises, logic is necessary to determine if a statement should be evaluated as true, false, or nonsense.

Logic operates on arguments. Arguments don't mean that people are fighting each other. It means someone has stated a conclusion or assertion, and backed it with at least one premise.

"All men are mortal. Socrates is a man. Therefore, Socrates is mortal." This is an example of an argument with a conclusion: "Socrates is mortal," and two premises, "All men are mortal" and "Socrates is a man." We don't need to determine if Socrates is man, or if all men are mortal. Our job is to determine *if the conclusions must be true if the premises are true*. If that's the case, as it is here, then we call the argument *valid*. If the premises are true, then we call the argument *sound*.

Making a distinction between the truth of the premises and the logical validity of the argument itself is one of the most fundamental skills in critical thinking. Often we focus on making our premises correct, neglecting the logical validity of our arguments, or focus on validity, but fail to justify our premises. People who wish to mislead us will often use arguments that are logically valid, but with false premises, or with true premises, but faulty logic. Separating the *form* and *content* of arguments is one of the first steps of critically thinking about any issue, and something that thinking about logic trains us to do.

Logical arguments can be broken into short statements with a format something like "If A, then B. A, therefore B." "If there is smoke, then there is a fire. There is smoke, therefore there is fire." If we wanted to analyze this statement, one method is the *counterexample*. Can we think of a circumstance where A is true, but B is not? If we can, then the statement is not valid. Are there circumstances where there may be smoke, but not fire? Since there are many circumstances where smoke can exist without fire, the statement is not valid. If the premises of an argument are themselves false, then the argument is said to be "unsound." It is important to remember that an argument must be both sound and valid: Both the content and the form must be correct. Formal logic is a rich and complex topic.

Fallacies

Formal fallacies are structural mistakes in the deductive reasoning in an argument. For example, "Some men are space aliens. Socrates is a man, therefore Socrates is a space alien." This argument is both unsound and invalid. The argument is invalid because the premises only state that *some* men are space aliens, not all, so it doesn't follow that Socrates must be a space alien.

Informal fallacies can happen with a deductively valid argument, but when the premises aren't handled correctly, or the argument isn't otherwise reasonable. Many arguments can be thought of as informally fallacious because they don't support their own conclusions, and make errors in reasoning.

These are usually called *logical fallacies*. These are not technically "*logical* fallacies" since the logic itself may be valid. It has been argued that informal fallacies have little use in teaching critical thinking because for a single fallacious passage, one can argue for or against multiple fallacies, and for some reasonable passages, logical fallacies might be applied (Hitchcock 2017). We see them as necessary but not sufficient for critical thought about argument and rhetoric.

Consider a television advertisement that takes the form "If sports stars agree on something, you should do it. Four out of five famous sports stars agree that you should buy Stan's tooth powder. Therefore you should buy Stan's tooth powder." There isn't anything logically wrong with this statement. However, a reasonable person would disagree with the premise "If a sports star tells you something, you should do it." This common kind of argument gets a special name, the *argument from authority*. Many similar classes of argument may be logically valid, but fail for other reasons.

While the argument from authority is questionable, often it is useful to look to people with expertise for guidance. It is very reasonable to value the opinion of an expert biologist over someone with no training on the question of vaccine safety.

As with the argument from authority, fallacies must be considered in context. They are useful tools for thinking about how people argue, but there are exceptions.

Some Fallacies

An *ad hominem* attack is a kind of argument against a person rather than their stated beliefs, premises, or conclusions (Van Vleet 2021). A physician might say to another, "Why should I listen to you? You went to a medical school that isn't even ranked in the top 60! I went to a school that's ranked 59, and yours was only ranked 62." The medical school someone went to isn't relevant to whether the course of treatment they propose is correct.

The *tu quoque* fallacy means "you too" and points out that the person you're arguing with is a hypocrite. For example: "How could you tell me not to put a fork in a socket, I saw you do just that last week!" Just because I stick forks in sockets doesn't mean that I'm wrong to point out to you that it isn't safe when you do it (my personal experiences may provide valuable insight into the outcome).

The *qui bono* fallacy asks "who benefits." Someone might reject all research funded by a corporation, because they believe that the financial motive of the research makes it unreliable. "I know that drug won't work to cure my sinus infection because Big Pharma makes money every time you buy an antibiotic. I'm going to spend my money on essential oils instead." Someone benefiting from an argument doesn't mean that the argument is wrong.

The *appeal to nature* fallacy happens when the quality of being "natural" is used as a selling point for an argument. "You should buy our product; it is more natural than another product." "You shouldn't do that, it's unnatural." Being natural (or less artificial) doesn't make an argument true, doesn't make a product better for health, and doesn't make a medical treatment better. There are many dangerous and unpleasant things in nature such as radioactivity, cyanide, and Nickelback. It is impossible to create a definition of nature that separates it from human actions.

The *appeal to tradition* is the use of tradition to argue that a method is superior to other methods (Michaud 2018). Because things have been done one way does not ensure that this is the best way to do things. "The best way to close this wound is with silk sutures, that's what I was taught in school, and it's what I've done for forty years." Tradition does mean new materials are worse. The *appeal to novelty* is the opposite.

The *bandwagon fallacy* describes the argument that because a belief or action is popular, it must be correct. "Normal people don't think the Earth is round. Why can't you just be normal?" More than one person can be wrong, or prone to the same errors in thinking. The popularity of an idea doesn't address whether it is true or not. If we count the number of times that masses have been wrong we would run out of real-number integers.

The *strawman* argument happens when rather than address someone else's true viewpoint, you create a false viewpoint that is easier to argue against. If someone has "put words into your mouth," you were likely experiencing a strawman argument. Suppose person A says, "If we impose a $0.005 per dollar sales tax in town we could afford more library services," and person B says, "I can't believe that you want to stifle business and expand big government." Person B hasn't responded to the actual proposal that person A made, but has responded as if they made an entirely different argument.

The *Texas sharpshooter*[1] fallacy refers to a way of overemphasizing positive data while deemphasizing disconfirming results. "A Texan wishes to prove his skill as a marksman, so he points his shotgun at a barn and fires, he then paints a target around the area with the most holes." The Texas sharpshooter fallacy occurs when someone selects only data that supports desired conclusions.

Hasty generalization is a fallacy that occurs when you draw conclusions about a large population given only a small sample size. Consider a situation in which you are a physician in a large hospital system in a city. In the last few days you have seen five cases of flu when typically this time of year you don't see any. You conclude that this will be an unusually bad year for flu cases.

The *sunk cost* fallacy occurs whenever the costs put into an endeavor take precedence over the potential future returns. People will often reinvest into failing endeavors because they focus on the amount of money, time, or energy already invested, rather than the potential for future rewards. "We've been doing this surgery for four hours already, if we stop without locating the tumor, all that time will be wasted—we had better continue." "I've been playing this slot machine for sixteen hours, it's bound to pay out." "I've been in this PhD program for five

---

1. The Texas sharpshooter fallacy and the illustrative anecdote that accompany it are widely used to discuss this logical fallacy, although the concept has been discussed in many forms over time. There is no evidence that it refers to any specific event or person. Some readers may object to such a term due to a history with gun violence or sensitivity to the mass shooting events of recent decades. Several other terms may be appropriate. We suggest that post-hoc clustering bias or retroactive precision fallacy are equally descriptive.

years; I can't quit now." The sunk cost fallacy is also known as the "gambler's fallacy."

The *post hoc ergo propter hoc* fallacy happens when correlation is confused with causation. When events happen at the same time, it is easy to believe that one event caused the other. "I stopped taking vitamin C for one day and I got a cold, so vitamin C must cure colds." Unlikely events often occur close in sequence to one another. For example, at times when ice cream sales are high, shark attacks are more common in the United States. Both events are activities that people are likely to participate in during warmer months, but one doesn't cause the other.

The *argument from ignorance* is a kind of fallacious argument people will use for a favored explanation of a phenomenon. "No one can explain what I saw in the sky, so it must have been aliens." An argument from ignorance occurs when the absence of a known explanation or ignorance of the known explanation for a phenomenon is used as evidence for a specific explanation. Not knowing the explanation of a phenomenon just means that "I don't know" is the current explanation.

The *middle ground fallacy* occurs when the middle ground between two opposing viewpoints is assumed to be the truth because it does not represent "either extreme." Being extreme does not make a view incorrect. Being in a middle position doesn't increase the likelihood that a viewpoint represents the truth. Consider a TV news program that decides that for every issue, both sides must be presented. When reporting on childhood measles vaccination, the program gives equal time to scientists explaining the weight of scientific evidence on the topic, and to someone who arrived at the anti-vaccination view primarily from social media. One viewpoint is much more strongly supported, but the news program has enforced an artificial middle ground and gave equal credence to unequal viewpoints.

A *false dilemma* occurs when a problem is presented as having only two possible solutions, as being a choice between only
two options, or as binary rather than lying on a continuum.
Binary constructions can eliminate choices that should not be
rejected a priori. "You're either for us or against your country.
It's as simple as that," is a line from the satirical novel *Catch-22*
that illustrates the absurdity of this way of viewing matters. A
romantic partner might say, "Either you'd greet me with more
enthusiasm or you don't love me," or a physician might say,
"It can only be heart disease or kidney failure."

A *shift in the burden of proof* occurs when someone makes a
claim, and then asks someone else to show that it *isn't* true.
The person making a claim is responsible to support the claim.
No one is responsible to refute it. It is possible to make infinite
unlikely claims and then ask others to do the work to show
which ones are not true. This will result in many untrue things
being believed until sufficient evidence exists to reject them.

Rhetorical fallacies are not quite logical fallacies because they
don't have the same structure, but use rhetoric to attempt to
convince someone of something. For example, the *appeal to
flattery* is a common rhetorical tactic where someone will use
flattery to persuade someone. "I can tell you're an intelligent
person, and that's why I'm offering you the chance to get in
on the ground floor of this offer. You're someone who can
appreciate what a good opportunity this is; all you have to do
is convince three more people to join our program and you'll
be making money!" Here the appeal to flattery is being used to
promote a multi-level marketing program, or pyramid scheme.

Conclusion

Logic is a powerful tool to help you avoid mistakes in thinking. If you learn to uncouple the form of the argument from

the truth of the premises, you can avoid many common mistakes in thinking. Learning about common logical fallacies can help you to recognize mistakes in thinking that are common, and avoid being persuaded by a bad argument. Thinking deeply about premises, validity, and conclusions can also help you to be a more persuasive thinker.

Summary

- Logic is a mechanism of understanding whether the conclusions of an argument flow from its premises, called validity.
- An argument must also be "sound," meaning that the premises are factually true.
- Logical fallacies are common mistakes in reasoning and argument that people make.
- Learning to uncouple soundness from validity and to recognize common logical fallacies can improve the way you reason through arguments and can help you understand the mistakes that you and others make

Exercises

One way to make sure that you aren't using logical fallacies in your own thinking is to practice using them deliberately. Pick a belief that you hold to be true and argue for it using logical fallacies. For example:

1. *Unicorns exist.* Multiple lines of evidence lead me to this conclusion. First, although I haven't seen one, I have viewed many TikTok videos where people recount their firsthand accounts of unicorn encounters. Some of these people occupy trusted positions in society such as firefighters or teachers. I feel very strongly that unicorns exist and it's important to go with your gut. Sometimes in life you just need to have faith in something. I have faith that unicorns exist. You can't prove that unicorns don't exist. You haven't been to every

forest in the world; maybe unicorns live in very remote areas away from humans, so they're hard to find!

Can you identify the logical fallacies used above? Can you write your own similar paragraph using logical fallacies?

2. Pick another belief you hold and write a paragraph explaining *why* you believe it to be true.

   a. Consider the arguments you used. Are any of them fallacious? Does your belief hold up if you don't continue to believe that argument?

   b. Select the remaining arguments. Can you think of *counterexamples* that might challenge your own belief? Counterexamples are examples of situations that, if they were true, would show an argument not to be sound. For example, if you argue that "all ducks can swim," a duck that cannot swim would be a counterexample.

Works Cited

Hitchcock, D. 2017. "Do the Fallacies Have a Place in the Teaching of Reasoning Skills or Critical Thinking?" In *On Reasoning and Argument: Essays in Informal Logic and on Critical Thinking*, edited by D. Hitchcock, 401–408. Springer International.

Michaud, N. 2018. "Appeal to Tradition." In *Bad Arguments: 100 of the Most Important Fallacies in Western Philosophy*, edited by Robert Arp, Steven Barbone, and Michael Bruce, 121–124. Preprint at https://doi .org/10.1002/9781119165811.ch19.

Van Vleet, J. E. 2021. *Informal Logical Fallacies: A Brief Guide*. Rowman & Littlefield.

5 Cultivating Competence: Expertise, Self-Assessment, and Professionalism

The whole problem with the world is that fools and fanatics are always so certain of themselves, and wiser people are so full of doubts.
—Bertrand Russell

Morgan is a new PhD student. Their first day as a PhD student is exciting! They gets to meet new people, make new friends, and feel as if they have arrived at a career that they've dreamt about for years. However, they start to notice the many accomplishments of their peers. Sometimes the peers appear to "get stuff" before they do, have worked at prestigious internships, or claim to be reading "Methods of Enzymology" at home for fun. Morgan starts to feel self-doubt, and wonders if they really belong. The classes get more difficult and they start to wonder if they have to work harder than their peers. They wonder if they lucked into their position, and if at some time everyone will realize that they're a fraud who doesn't actually belong in graduate school.

Morgan's experiences are near universal among graduate students and medical students. No one is willing to admit to one another that they feel like the other is smarter than them, and everyone suffers the same feelings. This feeling has been termed "impostor syndrome."

Impostor Phenomenon

Impostor phenomenon is a common feeling among students that they are unqualified or unable to work at the appropriate level for their fields, and they exhibit anxiety that they might be "exposed" as an impostor. It was initially thought to be exclusive to highly achieving women (Clance and Imes 1978); however, as the phenomenon has been studied, the range of those with impostor phenomenon has expanded to men, medical students (Henning et al. 1998), and is now thought to affect 70 percent of people (Gravois 2007).

Impostor phenomenon has been described as the product of a cycle of:

1. Not internalizing success and discounting positive feedback
2. Feeling doubt and anxiety
3. Needing to perform tasks to achieve
4. Anxiety leading to overpreparation or procrastination
5. Accomplishing tasks
6. Receiving positive feedback
7. Returning to a position of not internalizing success or positive feedback by attributing success only to hard work or luck. (Sakulku and Alexander 2011).

Although Morgan is as accomplished as the other students in their class, they still feel inadequate.

Various factors can contribute to the impostor phenomenon. Many students are coming from a school where they were the top of the class, and are not able to reconcile receiving imperfect grades for the first time in their lives. Likewise, many students are inexperienced with failure. When unreasonably high goals are not achieved, students feel as if they have failed even though they may have done well enough. Students may also feel guilt about accepting successes and praise.

Impostor phenomenon feelings are unhelpful for a few reasons:

1. Feeling anxious about a task doesn't accomplish the task.
2. The impostor phenomenon means that success cannot lead to happiness. If we reject our own successes, then when we achieve larger goals, we won't allow ourselves to be happy.
3. It is almost universal. If 70 percent of people experience it (and even more in medical or graduate school), then it cannot be an accurate reflection of your abilities. Not everyone can be an impostor.

One approach to combating the impostor phenomenon based on cognitive behavioral therapy is to recognize when it happens and replace the distorted thoughts with correct ones. Consider carrying a notebook, and when you notice an unhelpful thought write down three columns. Take a look at Morgan's notebook (see table 5.1).

Morgan has started to record their impostor syndrome thoughts and started to replace them with thoughts that are more helpful. This process takes time, as Morgan needs to keep reminding themself to recognize the distortions in their own thinking.

A phenomenon related to impostor syndrome is the arrival fallacy. The arrival fallacy is a tendency to believe that happiness depends on achieving certain goals in life, and that there will be subsequent feelings of emptiness after those goals are achieved. Accomplishing milestones rarely improves long-term happiness. People tend to overestimate the duration of their negative reactions to life events like a breakup or tenure denial that have significant short-term effects on happiness, but have little long-term effect (Gilbert et al. 1998).

Morgan finally makes it through graduate school all the way to the defense. It's a moment Morgan's been looking forward to for

Table 5.1

Thought	What's Wrong With It	Replacement
Everyone here is smarter than me.	Everyone went through the same process to get here. The committee looked into my background and decided that I'm qualified. I'm idealizing others while devaluing myself.	Other students have different life experiences; it's cool they were able to answer the question, but I wasn't. I deserve to be in this program because of my hard work, skills, and passion. The committee saw my potential.
I can't believe I got an 82 on that exam, even though I studied for weeks! I'll never succeed, I'm not good enough.	"Not good enough" is a permanent trait, and an exam is a single snapshot of how I performed this one time. It isn't all or nothing. I'm exaggerating.	I didn't do as well as I'd like. I'm going to go to office hours, ask questions, and adjust my study habits to be more effective. I can learn from this experience and improve.
I feel like I'm working ten times harder than everyone else. I can't relax. I can't watch TV. I can't take any breaks or I'll fall even further behind.	This is perfectionistic, and puts me under undue pressure. I need to set healthy expectations for myself so that I'm well rested and able to perform at my best. My self-worth shouldn't be tied into my productivity.	I've been working hard, but I can't go 100 percent all the time. I need some down time to rest. Taking an occasional break will help me perform better and improve my overall well-being.

six years. Morgan worked hard, struggled, published, and learned a lot during their time in graduate school. Morgan is proud of what they've accomplished, and confident that their postdoc will go well. The defense seems to go well.

Morgan steps out of the conference room, and waits while the committee deliberates on whether to grant a PhD. Finally, the committee steps out and congratulates them: "Dr. Morgan Clark." Morgan thanks the committee for their time. It doesn't feel right. Morgan feels empty, and doesn't know what they can look forward to anymore. Morgan feels alone, since no one is around to celebrate with them. Anxiety about the future creeps in, and Morgan feels unmoored from the goals and structure that have defined their life for the last six years.

The arrival fallacy is thought to happen because when we're focused on big goals like medical school or a PhD defense, we tend to see the end of these processes as landmarks that will change who we are and mark a new phase in our lives. However, the changes that happen in us during these processes happen over a long period of time. You don't become a physician all at once at the end of medical school when you recite an oath—it happens over the course of years of study and hard work. It helps to set reasonable expectations for how we'll feel when we reach milestones in our careers and lives.

Professionalism

A profession is a discipline that meets certain requirements such as a code of ethics, a governing body, specialized education, autonomy, and licensing. In the United States, for example, a lawyer is bound to follow a code of legal ethics, is governed by a state bar, must complete a legal education at a law school, is licensed to practice law. Likewise, professions gain autonomy in the sense that only other lawyers may provide input into their practice of law.

Many fields have undergone professionalization, including surveying, medicine, and accounting. Medicine in the United States is governed by the American Medical Association (AMA) and American Osteopathic Association (AOA), and similarly has a specialized education system, codified ethics, autonomy, and licensing; thus, it is a profession, and medical school is called a "professional school."

Science has not been professionalized, and could and should never be professionalized. Although there is a job sometimes called "professional scientist," this usually just means someone who is paid to do science. Part of what makes science work is its openness to changing beliefs in light of new evidence, and considering ideas from any source. Closing off avenues of evidence due to being from the wrong source would violate that principle of the scientific attitude. For science to continue, then, it must be true that *anyone* can contribute to research, have a new idea, or collect new data.

Professionalization can have benefits, such as standardization, which assures clients/patients of the quality of service provided, and financial benefits, by limiting the number of members. However, professions can potentially exclude qualified people from working in them, operate monopolistically, or limit creativity. For example, at present in the United States there are more medical graduates than there are residency slots due to federal caps that have existed since 1997. The AMA had a role in supporting the budget agreement that established those caps. The AMA denies being aware of the long-term implications of those caps, but at the time feared a surplus of physicians.

Another meaning of professionalism is the set of norms that are enforced or expected in certain workplaces such as punctuality, dress codes, and decorum. Some of these make sense. It is reasonable to enforce norms for behavior such as solving problems by discussion rather than shouting matches. Likewise, most workplaces enforce norms for dress such as cleanliness and covering areas of the body deemed to be culturally taboo.

Standards of professionalism differ between schools, and there are no universal standards that are agreed upon between medical schools. However, some ideas might be shared between many institutions, such as:

- Maintain good personal hygiene.
- Use respectful language when talking to (or about) colleagues, teachers, allied health providers, patients, or patient families.
- Listen and consider others concerns.
- Protect patients' privacy.
- Remain punctual.
- Do what you say you will do.

There can be negative consequences of this kind of "professionalism" as well, such as enforcing norms counter to your interests. Professionalism could be used as an excuse in fields dominated by men to exclude women, or to enforce political/ ideological views. Professional standards can also act as shibboleths of in-group or out-group membership.

The "white coat" or lab coat is an example of how a professional standard can become detached from practical utility. Lab coats were initially worn by scientists to prevent experimental contamination and as personal protective equipment. Physicians began wearing them to show the connection between medicine and science; however, a growing body of evidence shows that physician lab coats can carry bacteria from patient to patient (John et al. 2017), and short sleeves and scrubs are now recommended. Nonetheless, "whitecoat" ceremonies are still held by 97 percent of medical schools as symbolic of the profession.

The Dunning–Kruger Effect

If someone experiencing impostor syndrome is making an error where they assume they are less competent than they are, then someone experiencing the *Dunning–Kruger effect* is doing the opposite. They are unaware of the limitations of their own competency.

The Dunning–Kruger effect is often explained as "the less you know about a topic, the more you believe yourself to be an expert on the topic." However, this explanation isn't exactly what Dunning and Kruger found in their original paper (Kruger and Dunning 1999). Dunning and Kruger's paper "Unskilled and Unaware of It: How Difficulties in Recognizing One's Own Incompetence Lead to Inflated Self-Assessments" looked at people with differing skill sets in different cognitive tasks, grammar, logical reasoning, and humor. The participants were assessed in their skills, and then asked to assess their own ability compared with others. Competent individuals tended to rate themselves highly on skill. Incompetent individuals rated themselves less highly than competent individuals, but higher than their own actual skill level. When individuals were trained in the tasks, their self-assessments came into better alignment with their skills.

We may not be aware of precisely how little we know in a domain of knowledge while we are unfamiliar with it. Inaccurate self-assessment can have significant impacts in science and medicine. Medical students are likely to overestimate their own communications skills (Gude et al. 2017). Physicians in the lowest quartile tended to rate themselves 30–40 percentile ranks higher than peers (Violato and Lockyer 2006). Likewise, medical students will rate their skills higher than expert reviewers at tasks like interviewing standardized patients (Hodges et al. 2001).

The inability to self-assess may have potentially dangerous consequences, leading to physicians who make medical errors, communicate ineffectively with patients, do not seek additional professional training, and undermine trust in science and public health by promoting bogus treatments. Likewise, scientists who do poorly at self-assessment may produce lower quality research, fail to seek professional development, or refuse to accept the contributions of skilled peers.

We can all benefit from outside evaluation and feedback. If we develop a sense of humility and a culture of continuous learning and peer review, we can mitigate some of the dangers of the Dunning–Kruger effect. Leaders can model this sense of humility and openness to feedback, can regularly seek honest peer and expert feedback on one's performance and knowledge, and can seek opportunities for continued learning. Medical and scientific learners can benefit from interactions with those outside their immediate fields and those in allied health professions, and can benefit from mentoring junior colleagues with honest assessments.

Summary

- At some point or other many people feel inadequate in their positions or feel as if others are making more progress, and often these feelings are misplaced.
- However, people often overestimate their own abilities, especially if they are unfamiliar with a field.

- Professionalism is both a set of institutional behavior rules and a system for controlling access to and advocating for a field.

Exercises

1. Fill out the table below with some of your own thoughts that might relate to impostor syndrome. Identify problems with your own thinking, and rewrite the thoughts in a more positive way:

Thought	What's Wrong With It	Replacement

2. Make a list of your positive accomplishments, qualifications, and positive feedback you've received from others. Share that list with a peer and see if they can add any items to your list.

3. Write a paragraph describing how you view professional behavior at your institution. What rules are in place? Are they written down or unspoken? What reasons exist for the rules?

4. Work with a peer who knows you well and can provide honest feedback. Write down on a scale of 1–10 how you evaluate your own skill on the topics below, and compare with your peer. Do you overestimate your own ability?

   a. Communication
   b. Problem-solving
   c. Critical thinking

5. Create an inventory of your own skills and knowledge as well as an assessment of your own abilities in it. Keep this list updated throughout your education as you learn new topics. Does your self-assessment change?

6. Write a reflective paragraph about your own goals. What do you want to learn in school? Do you see graduation as an end point or a beginning in your education?

Works Cited

Clance, P. R., and S. A. Imes. 1978. "The Imposter Phenomenon in High Achieving Women: Dynamics and Therapeutic Intervention." *Psychotherapy* 15: 241–247.

Gilbert, D. T., E. C. Pinel, T. D. Wilson, S. J. Blumberg, and T. P. Wheatley. 1998. "Immune Neglect: A Source of Durability Bias in Affective Forecasting." *Journal of Personality and Social Psychology* 75: 617–638.

Gravois, J. 2007. "You're Not Fooling Anyone." *Chronicle of Higher Education* 54.

Gude, T., A. Finset, T. Anvik, et al. 2017. "Do Medical Students and Young Physicians Assess Reliably Their Self-Efficacy Regarding Communication Skills? A Prospective Study from End of Medical School Until End of Internship." *BMC Medical Education* 17: 107.

Henning, K., S. Ey, and D. Shaw. 1998. "Perfectionism, the Impostor Phenomenon and Psychological Adjustment in Medical, Dental, Nursing and Pharmacy Students." *Medical Education* 32: 456–464.

Hodges, B., G. Regehr, and D. Martin. 2001. "Difficulties in Recognizing One's Own Incompetence: Novice Physicians Who Are Unskilled and Unaware of It." *Academic Medicine* 76: S87–S89.

John, A., H. Alhmidi, M. Gonzalez-Orta, J. Cadnum, and C. J. Donskey. 2017. "Bare Below the Elbows: A Randomized Trial to Determine Whether Wearing Short-Sleeved Coats Reduces the Risk for Pathogen Transmission." *Open Forum Infectious Diseases* 4: S34–S34.

Kruger, J., and D. Dunning. 1999. "Unskilled and Unaware of It: How Difficulties in Recognizing One's Own Incompetence Lead to Inflated Self-Assessments." *Journal of Personality and Social Psychology* 77: 1121–1134.

Sakulku, J., and J. Alexander. 2011. "The Impostor Phenomenon." *Journal of Behavioral Science* 6: 75–97.

Violato, C., and J. Lockyer. 2006. "Self and Peer Assessment of Pediatricians, Psychiatrists and Medicine Specialists: Implications for Self-Directed Learning." *Advances in Health Sciences Education: Theory and Practice* 11: 235–244.

6   Thinking with Numbers 1: Probability in Science and Medicine

Probability is expectation founded upon partial knowledge. A perfect acquaintance with all the circumstances affecting the occurrence of an event would change expectation into certainty, and leave neither room nor demand for a theory of probabilities.

—George Boole

Probability

Probability studies the likelihood of events, and has myriad uses in science and medicine. Punnett squares are used for calculating the likelihood of a trait being passed between generations. Epidemiological models use probability to predict the spread of infections. Probability also describes important chemical and physical properties of matter. In medicine, probability is necessary to correctly interpret test results, estimate survival likelihoods of procedures, and determine if tests should be performed, and what diagnoses are to be considered. Probabilities are often counter to our intuitions and expectations, leading to misunderstandings, which can have serious consequences.

We might assume that events are independent when they are not, such as assuming that your probability of inheriting a genetic disease is independent of your parent's genotype. It is common to assume events are not independent when they are, such as believing that prior exposures to a virus without getting sick mean that you cannot get sick again, when in fact each exposure is an independent event with its own probability. We might look at data and fail to believe it because its "clumpiness" feels wrong. For example, imagine a clinical trial where participants are assigned to a placebo or experimental group at random, and the sequence PPPPPPPPEEE (where P is placebo and E is experimental) comes up.

This has the same probability as the sequence PEEPEPEEEPE, although our intuition suggests that the second is more random. Measurements of biological values such as blood pressure are often taken once at a medical visit and written down as a single value. However, our blood pressure varies throughout the day due to circadian rhythms, stress levels, diet, exercise, and sympathetic system tone. An individual measurement might significantly vary from this mean. An initial measurement of 160/105 mmHg may be followed by a series of measurements that cluster around 120/80 mmHg, in a process known as *regression to the mean*. Understanding probability can help us avoid the pitfalls of our own intuition.

Common Mistakes in Thinking About Probability

The Fallacy of Conjunction

Linda is 31-years-old, single, outspoken, and very bright. She majored in philosophy. As a student she was deeply concerned with issues of discrimination and social justice, and also participated in anti-nuclear demonstrations. Which is more probable?

1. Linda is a bank teller.
2. Linda is a bank teller and is active in the feminist movement. (Brogaard 2024)

Surveys indicate that people are more likely to decide that option 2 is more probable (Tversky and Kahneman 1983). However, one of the basic rules of probability, *conjunction*, teaches us that this is not true. The probability that A is true is always higher than the probability that A and B are both true. Say that the odds that Linda is active in the feminist movement are 99 percent, and the odds that she is a bank teller are 50 percent. To estimate the probability of both being true, we would have to calculate 0.50×0.99, which is 0.495, less than 0.50 (also see chapter 3).

Errors of conjunction are common, and can lead to incorrect conclusions. A physician might assume that a patient who is obese does not exercise, for example. This error in probabilistic thinking is known as the *fallacy of conjunction*. There are two important risks from the fallacy of conjunction:

1. *Stereotyping*, or assuming that because one person (or situation) shares some properties with another person (or situation), they share all other properties. The example problem above plays into stereotypes of someone who might be active in the feminist movement.

2. *False attribution of cause.* A scenario where this fallacy might play a role is the consideration of a medication. You may have a patient you placed on lisinopril for high blood pressure. Nonetheless, the patient has a stroke. "This medicine doesn't work!" you say, and never prescribe lisinopril again. However, the probability of stroke given that the patient takes lisinopril is lower than the probability of lisinopril with no stroke.

Sensitive and Specific

A test is developed for a rare disease that affects one person in your town. The test is very *sensitive*, and has a *specificity* of about 95 percent. As it happens, your town is home to 1,000 people, and you decide to test each one for the disease. How many tests return positive? How many are a true positive, and how many are a false positive?

You would expect fifty-one positive tests, and only one of those (< 2%, or 1/51) actually has the disease. Why is that? Every test has a sensitivity and a specificity. The sensitivity is also known as the *true positive rate*. It is the likelihood of having the condition if the test is positive. The specificity is known as the *true negative rate*. It is the likelihood that the test will return negative if you do not have the condition. In this case, the test is 95 percent likely to return a negative result if you do not have the disease. However, because you are testing a large number of people, 5 percent of 1,000 is 50. Fifty is far more than the true number of people with the disease.

False positives, false negatives, and *overtesting* are common but avoidable. For example, let's propose that the treatment for the disease is a surgery with a 90 percent survival rate. If we test everyone, and operate on everyone who tested positive, five people will die who were false positives, instead of the one who had the disease. *Overtesting* can potentially result in more harm than good. In 2016 the US Preventive Services Task Force recommended raising the age at which mammograms are recommended every two years to age fifty (Siu and US Preventive Services Task Force 2016). Age is the most important factor in breast cancer risk. Treating as a result of false positives carries its own risks. However, as age advances and the risk of a deadly cancer increases, the benefits of screening begin to outweigh the risks. These guidelines have very recently been reversed as more women in their forties began developing breast cancer, and the balance of risks once again shifted.

Understanding true and false positive rates is also important for scenarios such as checkpoints for drivers under the influence, and security screenings, which can create legal problems for more innocent people than guilty. In science, the sensitivity and specificity of tests is important for the interpretation of research results.

Testing too many hypotheses can create errors due to false positive rates. For example, imagine a technology with a 95 percent specificity that tests protein–protein binding, and allows you to test a protein against the entire transcriptome of a cell. Dozens of significant protein–protein "interactions" might show up in this analysis, but only a handful will be "real."

We can limit our tests to only cases that we have a good reason to suspect are going to be positive. Suppose we learn that the disease has a common symptom of a runny nose. On a given day, only five people out of 1,000 have a runny nose, so only those people are tested. The likely outcome is only the person with the disease testing positive. Performing mammograms only on those likely to have lethal breast cancer reduces harm from false positives; performing breathalyzer tests only on those who are driving erratically reduces false positives; and testing for interactions only between proteins hypothesized to interact reduces false positives. It is important to have a strong *prior probability* before running a test.

The Clumpiness of Data

Imagine that you construct a radio tower on your property and on January 1 it is struck by lightning. Based on the number of lightning strikes per year, you estimate that the likelihood of the tower being stuck on any given day is 1/100, or 0.01. On January 2 the tower was struck by lightning again. What are the odds?!

When things happen right after one another, it's easy not to see it as a coincidence, but in reality stochastic events often appear "clumpy." If you flip a coin enough times, you will see that there are short runs where there will be multiple heads or tails in a row. Flip coins long enough and you may find five tails in a row. If you continue to flip, you may eventually get six tails in a row, and so on.

Back to the tower. The likelihood that the next lightning strike will be on January 2 are 1/100. What is the likelihood that the next lightning strike is on January 3? Our first thought is probably 1/100. However, for a strike on January 3 to be the *next* strike, lightning has to *not* strike on January 2. We know that the likelihood lightning won't strike on January 2 is 99/100. Therefore, the odds that the next lightning strike occurs on January 3 are 0.99×0.01, or slightly lower than 0.01. The odds for January 4 are $0.99 \times 0.99 \times 0.01$, or even slightly lower. So the most likely day for the next lightning strike will be the day following the most recent strike, even if the likelihood of a strike on a given day is 1/100.

Humans tend not to expect data to be clumpy, and tend to perform poorly at simulating it when trying to recreate stochastic data. This tendency can produce patterns that have been used to identify cheating teachers in Chicago public schools (Sheth 2017), as teachers faked test results in a way that didn't exhibit the expected degree of clumpiness. This tendency not to expect clumpiness can get us into trouble when noticing patterns. If we notice a rare event, we might think another similar rare event is *less* likely to occur in proximity. For example, imagine you're working in a hospital and you encounter a patient with all of the symptoms of a rare disease, von Recklinghousen's disease. Later that night you see another patient with the same symptoms. During all of your training you've heard the phrase "When you hear hoofbeats, don't look for zebras," meaning don't look for rare diagnoses when a less-rare diagnosis might fit. In fact, the odds of each diagnosis are independent, and should be assessed independently each time.

In a broader sense, the tendency to see clumpiness as patterns can cause other problems to thinking, such as reinforcing conspiratorial thinking or reinforcing weak hypotheses. Any time we think, "This *can't* be a coincidence; what are the odds?," we might be falling prey to this kind of thinking.

Bayeswatch

In medicine or science we rarely have fixed likelihoods for our hypotheses or diagnoses. More often, we need to walk through a process with our beliefs, updating them as new evidence becomes available. You might start with a list of possible diagnoses for a patient based on symptoms, and you might order a test, and update the list of likely diseases based on the outcome. You might order additional tests, and further update your beliefs based on those results. Likewise, you might continually update your beliefs about the explanation for data you generate in the lab, based on the results of additional tests.

This natural updating of beliefs in the light of new evidence is formalized by *Bayes' theorem*, which can be written in common language as something like "Instead of discarding your past beliefs with each new datum, update your beliefs." Bayes' theorem gives tools for formalizing how the likelihood of beliefs changes with new evidence. It can also explain a bit about how people see the world and can be used for critical thinking. Often, two people will observe the same event and walk away from it with two completely different interpretations. How can this happen if only one event occurred, and neither person is either lying or hallucinating? Both people observed the event and updated their views, but both had different prior beliefs that colored their interpretations.

Bayes' theorem reflects how we test hypotheses and ideas about the world in general. The problem of induction means that we can't test an idea and prove it to be true. We can, however, test ideas and increase the likelihood we assign to those ideas over time as we fail to falsify them. As new evidence becomes available, we might rely less on the initial evidence that led to our earliest beliefs. If we say, "The patient's chief complaint is a cough; I think a cold is the most likely diagnosis," and a later test shows a blood glucose of 612 mg/dL, we should think, "The patient very likely has diabetes," and not "It doesn't seem likely that diabetes would cause a cough; it's probably just a cold."

Odds Ratio

The *odds ratio* is a metric commonly reported in clinical trials that gets at the association between two statuses. For example, say you want to understand a relationship between use of an inhaler and developing COVID-19 in a given month. You select a group of 1,000 people who don't use inhalers and 1,000 who do, and at the end of the month you test each group for COVID-19 antibodies. In the first group, 140 test positive. In the second, 100 test positive. The odds are 140/1000 in the no inhaler group, and 100/1000 in the inhaler group. These can be written as 0.14 and 0.1, respectively. The ratio 0.1/0.14=0.71. An odds ratio of 1 indicates no association, while a number above 1 indicates a positive association and a number below 1 a negative association. In this hypothetical example, the researchers would find a weak negative association between inhaler use and developing COVID-19.

Absolute and Relative Risk

Confusion between *absolute risk* and *relative risk* can lead to misleading interpretations. The absolute risk of a disease is the likelihood of the disease occurring to anyone in the population. A headline might say, "Eating a daily chocolate bar reduces the odds of glioma by 50 percent." This may suggest to you that you should devote 300 calories of your daily caloric intake to chocolate. However, this headline is reporting a relative risk reduction. The absolute risk of developing a glioma in your lifetime is perhaps only 4/1000 (NIH 2014). Eating a chocolate bar (fictionally) brings that risk down from 0.4 to 0.2 percent. The downsides of getting 10 percent of your daily caloric intake from candy may exceed the potential benefits. The absolute risk is an individual's risk of developing a condition in a population. The relative risk reduction is the fractional change in risk given a particular treatment.

Conclusion

This chapter is intended to explore your thinking about probability. There's a lot more to know about it, and learning the formal rules of probability from a more mathematical perspective can help you to better understand situations you might encounter in both science and medicine. This would allow you to make better decisions based on the science presented. There are a variety of texts that can help with this, such as OpenStax Introductory Statistics, which is available online for free.

Summary

- The probability of two independent events is always less than either event on its own.
- Tests can have false positives and false negatives. It is important to think about the false positive and false negative rates to determine criteria for when tests should be run.
- Random (stochastic) data tends to be clumpy; sometimes events can occur close together in time without being related.
- Instead of replacing our beliefs with each new piece of evidence, we can use new pieces of evidence to update our prior beliefs.

Exercises

1. Write a reflective paragraph describing how you would counsel a patient who wishes to have a mammogram despite being five years younger than the recommended age. What do you tell her? Do you attempt to dissuade her while respecting her autonomy? How does your answer change if she has a family history of breast cancer?

2. Search on the internet for the difference between "absolute risk" and "relative risk." Two drugs are compared in a clinical trial. One thousand participants receive either drug A or drug B. What are the relative risk and absolute risk reduction between drug A and drug B if 100 people in group A and 75 people in group B experience myocardial infarction?

Works Cited

Brogaard, B. 2024. "Linda the Bank Teller Case Revisited." *Psychology Today*.

NIH, National Cancer Institute. 2014. "Previous Version: SEER Cancer Statistics Review, 1975–2011." December 17. https://seer.cancer.gov /archive/csr/1975_2011/.

Sheth, A. 2017. "How Data Analysis Helped Uncover the 'Cheating Teachers' in Chicago Public Schools." *Technology Titbits.* https://medium.com/bloombench/how-data-analysis-helped-uncover-the-cheating-teachers-in-chicago-public-schools-ff1a52d3e00a.

Siu, A. L., and U.S. Preventive Services Task Force. 2016. "Screening for Breast Cancer: U.S. Preventive Services Task Force Recommendation Statement." *Annals of Internal Medicine* 164: 279–296.

Tversky, A., and D. Kahneman. 1983. "Probability, Representativeness, and the Conjunction Fallacy." *Psychological Review* 90: 293–315.

7 From Curious to Hypothesis

> Ideas, like ghosts . . . must be spoken to a little before they will
> explain themselves.
>
> —Charles Dickens, *Dombey and Son*

*Alice is a new graduate student who has picked her lab and just
started trying to learn from the literature in her chosen field. Her advisor wants her to come up with her own project, but she's never done
that before. Her instructions were that she needed to develop a project
around a testable hypothesis. She tells her advisor, Dr. Jones, that
she "wants to explore how cytoskeletal proteins interact with each
other." Dr. Jones tells her that this is a nice idea, but it isn't a testable
hypothesis, and that Alice should try again. Alice shows Dr. Jones the
hypothesis: "Interactions between cytoskeletal proteins cause cancer."
Dr. Jones says, "You're definitely getting closer, but which proteins?
Which cancer? How would you measure it? Is it consistent with the
literature? Take another day to think and try again."*

All hypotheses aren't made equal. Two hypotheses might be
formulated to explain the same phenomenon, but one might be
weak while another is strong. Developing a hypothesis requires
creativity, knowledge of a field, and enough experience to know
what will and won't be testable. In this chapter we'll discuss

what makes a hypothesis strong, and how we can think about hypotheses as we develop them.

As we discussed in chapter 1, the test for what makes a hypothesis scientific is falsifiability. However, a hypothesis can be falsifiable, but not very good for other reasons. We need more measures to distinguish how "strong" a hypothesis is. A hypothesis might be technically falsifiable, but also "weak" or uninteresting, in which case it is not worthwhile to examine.

Strong hypotheses are testable, specific, and measurable; are logically consistent without internal contradiction; are formulated so as to be open to revision; are based on empirical observation; make a prediction about a future event that can then be tested; explain many phenomena and aren't limited to a few; and are replicable so that different people running tests of the hypothesis should achieve the same result. We'll explore what each of these ideas mean.

Testable

A hypothesis that isn't testable isn't scientific because no science can be done on it. Astronomer Carl Sagan offered an example, by asking us to imagine the hypothesis that there is a dragon in his garage. Normally, this idea could be tested, but in this case it has been modified in such a way that it is more difficult to test. The dragon cannot be seen, touched, heard, smelled, or in any way sensed.

"Now, what's the difference between an invisible, incorporeal, floating dragon who spits heatless fire and no dragon at all? If there's no way to disprove my contention, no conceivable experiment that would count against it, what does it mean to say that my dragon exists?" (Sagan 2011). In Sagan's analogy, the dragon is a hypothesis that isn't testable. It cannot be seen, cannot be touched or sensed in any way, and no device can be devised to measure its effects. You cannot write a statement like "if the dragon is real, then I would expect . . . ," because the dragon has no expected effects on the world. Whether the dragon is real is therefore not a question that science can address, and not a very interesting question because the world we live in is the same whether the dragon exists or not.

There are limited resources in science, and we can't test every hypothesis we might have. This is why we also want our hypotheses to be strong.

Internal Consistency

A strong hypothesis should be internally consistent; that is, it should not contradict itself. The hypothesis "All hats are chairs" is self-contradictory, because hats and chairs are different things. Likewise, "No mammals can fly; bats are mammals" is inconsistent because bats belong to the category "mammals" and can fly. A hypothesis that says, "Smoking increases the risk for cancers; smoking is good for our health" would also be internally inconsistent.

However, some widely accepted hypotheses are internally inconsistent. For example, "Light is both a wave and a particle" is supported by a large amount of empirical evidence of light behaving like a wave in some circumstances and like a particle in other circumstances. Nature cannot lie, so wave–particle duality must be true. Most likely, the traditional concepts of "wave" and "particle" as distinct categories are flawed. Despite rare exceptions that require us to expand the way we think about words and categories like the example with light above, an internally inconsistent hypothesis is generally a weak hypothesis.

Openness to Revision

A hypothesis that is not open to revision is weak because no test could potentially show it to be wrong. An investigator who refuses to revise a hypothesis or accept that it has been falsified risks continuing to hold onto beliefs that are no longer likely to be true.

The most famous example of a lack of openness to new ideas is the house arrest of Galileo Galilei. Galileo had made astronomical observations that supported a heliocentric model of the solar system in which the Earth revolved around the sun. At the time, the heliocentric model was deemed "heretical," because the belief that the Earth was the center of the universe was considered to be Church doctrine. Authorities banned Galileo's book and he was placed in house arrest for the rest of his life. Those in the Church weren't scientists, but their ideas about the universe were still hypotheses that they refused to revise (until hundreds of years later, in 1992).

Another example is Joseph Lister, a nineteenth-century surgeon who discovered that wounds treated with a dilute solution of phenol (which we now understand was acting as an antiseptic) were less likely to develop gangrene, and that surgeries carried out with instruments and gloves that had been treated with phenol were less likely to cause deadly infections. The medical community of the time was initially reluctant to accept his idea, partially because the theoretical support—the germ theory of disease—was not yet accepted. Other ideas about the origins of disease such as terrain theory, the miasma theory, and ideas about spontaneous generation of bacteria were widely held. How could killing all the bacteria on a surgical instrument help prevent an infection if new bacteria would spontaneously generate in the wound anyway? Had Lister's critics done thorough experiments they would have come to similar conclusions as Lister: Chemical treatment of surgical instruments to kill bacteria reduced infections.

Most examples of failing to revise a hypothesis aren't as dramatic as these; however, it is important to be willing to abandon hypotheses when they are no longer tenable. Often people fail to do this and will instead default to an *ad hoc rescue*. A researcher might modify the sample size in a population to obtain different results, add additional variables to adapt to inconsistencies, downplay results not favorable to their hypothesis, or change analysis techniques after the fact so their hypothesis appears more likely.

A well-known example of ad hoc rescue are the great lengths that physicists went through to hold onto the idea of light as a wavelike substance that propagated through a medium of "luminiferous ether." When experiments didn't show the expected effects of luminiferous ether, some went as far as suggesting that the ether had altered the experimental equipment (Hunt 2012). The luminiferous ether was later rejected as a hypothesis. An ad hoc hypothesis is often used to explain away unexpected results, or results found to be inconvenient to favored ideas, so that existing theories do not need to be modified.

If the intent of a hypothesis was to modify a preexisting hypothesis so that it can "remain true" despite contradictory results, then the hypothesis is ad hoc. This kind of hypothesis rescue isn't always bad. Some hypotheses require refinement of this sort; however, too much ad hoc patching may be a sign that a hypothesis is weak and should be discarded.

Formulated Based on Empirical Observation

It is important to formulate hypotheses based on empirical observation because these hypotheses are based in the real world and lend themselves to testing. Hypotheses based on moral beliefs, intuitions, or opinions are more likely to be difficult or impossible to test, and unlikely to accurately reflect the world. That doesn't mean that a hypothesis formulated this way *can't* be true, but it does mean that this method of inquiry is likely to be less fruitful, and may make us more prone to investigative biases.

Because ultimately every hypothesis must be tested by empirical observation, hypotheses based on empirical observation are more likely to be accurate and reliable.

Explanatory Power

A hypothesis that explains only a limited number of observations isn't as useful a model as one that explains that as well as more observations. Generally speaking, the more a hypothesis can explain, the stronger it is.

A hypothesis with a lot of explanatory power can explain a great deal of data. A hypothesis with low explanatory power is one that, if true, would explain only a small amount of data. Say you are studying the behavior of gases in a vacuum. The hypothesis that oxygen gas will expand when heated explains less than the hypothesis that all gases will expand when heated. The hypothesis that addresses all gases includes the hypothesis that oxygen gas will expand when heated. If the hypothesis that explains more is true, then the less explanatory hypothesis is also true. That means that between the two hypotheses, the one with greater explanatory power is generally the stronger one.

Predictive Power

Consider that a scientist might adjust a pseudoscientific hypothesis to appear to encompass past data, but that the hypothesis might fail when used to predict a new outcome. The predictive power of a hypothesis is another characteristic of a hypothesis that makes it strong.

Charles Darwin observed the orchid *Angraecum sesquipedale*. This orchid has an unusually long spur. Darwin knew that other orchids had spurs with a length that correlated with the length of hawk moths that pollinated the orchids. Darwin hypothesized how such a spur might have evolved. He imagined a "kind of arms race" resulting in "ever longer spurs and hawk-moths with ever longer tongues" (Fay and Chase 2009). And "in Madagascar there must be moths with proboscides capable of extension to a length of between ten or eleven inches." His hypothesis led to a prediction: a secondary testable hypothesis, that a hawk moth had coevolved to pollinate the orchid *Angraecum sesquipedale*, and that it must have an eleven-inch proboscis. The pollinating moth was not discovered until twenty-one years after Darwin's death, and the moth was observed visiting the orchid in 1992, 130 years after Darwin made this prediction. Darwin's hypothesis had strong predictive power, predicting the outcome of an experiment that would not be carried out for more than a century. An alternative hypothesis such as "the flower's long spur was caused by supernatural intervention" doesn't offer any predictions that can be tested.

Replicability

Strong hypotheses are replicable. This means that different researchers, in different labs, with different equipment, in different locations should be able to get the same result by running the same experiments with the same controls.

Currently, a number of scientific fields are going through replication crises. Results that were published in the scientific literature were retested, and the results did not match those found by the original researchers. If the results of one team do not match the results of another, that calls into question the initial interpretation of the results and subsequent work based on it. Failure of replication reduces the confidence in results. The most famous replication crisis is in psychology. In 2015 the Open Science Collaboration published a study examining 100 studies published in 2008 (Open Science Collaboration 2015), some highly cited. Ninety-seven percent of the original studies had significant results, but only 36 percent of the follow-up replication studies did.

Similar replication crises have occurred in other areas of science such as cancer biology (Mullard 2021). These replication crises might be consequences of unintended incentives in scientific research, such as high pressure to publish research, poor education in statistical methods, and journals that accept small sample sizes. Since replicability is necessary for science to be trustworthy, a hypothesis that is designed to be testable under many circumstances is stronger than a hypothesis that requires special circumstances that might not be easily replicated.

Meeting Standards

Having a treatment or diagnostic tool enter the practice of medicine happens often, and ideally it happens through those ideas meeting certain criteria. Ideas that fail to meet these criteria should eventually be rejected. These criteria can change as we learn more about what works in medicine; however, some are widely agreed on.

Scientific publishing allows for ideas to be exchanged, subjected to peer review, and subjected to the scrutiny of other scientists. Peer review isn't a guarantee of quality, as many papers have made it past peer review that were later shown to be wrong in conclusion or method, but it can eliminate some problems.

Openness to criticism is necessary for any idea undergoing scientific scrutiny. If an idea cannot be critically examined and is not open to potentially being rejected, then it cannot stand on its own merits. Treatments that reject criticism as being "from outside the profession" or that refuse testing cannot be verified. Openness to criticism of ideas is another of the features of the scientific attitude that carries over to medicine.

Someone proposing a treatment is the one who must demonstrate that it works. Refusing to accept this burden of proof is a red flag. Statements like "Prove that it doesn't work" or "You haven't shown that my treatment isn't effective" is a *shift in the burden of proof*, a kind of logical fallacy. Critical thinkers start by assuming the null hypothesis, that a treatment or diagnostic tool will not work, and attempt to falsify that hypothesis.

Those honestly proposing new methods are self-skeptical. They look for the potential flaws in their own reasoning, possible confounding factors in their studies, and reasons to reject their own hypotheses. Self-skepticism is a part of the scientific attitude because it's easier to fool ourselves than anyone else. Scientists and critical thinkers rarely make bold statements of truth without qualifiers. To a lay audience, this can appear to be a lack of confidence in results, but in reality this is an expression of self-skepticism, a necessary quality for anyone asserting ideas.

Conclusion

Alice returns to Dr. Jones's office one final time with her new hypothesis: "A common polymorphism in the gene coding for beta-spectrin will lead to increased proliferation of HEK293 of CHO cells transfected with it." Dr. Jones says: "Good work, this is finally a testable hypothesis. Now, before we start doing experiments, let's figure out what statistical tests we're going to run."

Summary

- Hypotheses can be strong or weak.
- Strong hypotheses are testable, are specific, have predictive power, have explanatory power, are open to revision, are internally consistent, are replicable, and are formulated based on empirical observation.
- It is useful to invest in developing stronger hypotheses.

Exercises

1. Consider each of the following hypotheses. Are they strong or weak? Why?
   a. Eating garlic will cure a cold.
   b. All people who wear glasses are intelligent.
   c. Cheese is the secret to happiness.
   d. Money grows on trees.
   e. Cars can fly.
   f. All dogs are cats.

2. Find the database file MTCARS online. Download it and make a hypothesis about two of the variables in the data set.
   a. How will you test this hypothesis?
   b. Having tested this hypothesis, was your hypothesis supported or falsified?
   c. Share your hypothesis with a peer for feedback.

Works Cited

Fay, M. F., and M. W. Chase. 2009. "Orchid Biology: From Linnaeus via Darwin to the 21st Century." *Annals of Botany* 104: 359–364.

Hunt, J. C. 2012. "On Ad Hoc Hypotheses." *Philosophy of Science* 79: 1–14.

Mullard, A. 2021. "Half of Top Cancer Studies Fail High-Profile Reproducibility Effort." *Nature*, December 9. http://dx.doi.org/10.1038/d41586-021-03691-0.

Open Science Collaboration. 2015. "Psychology: Estimating the Reproducibility of Psychological Science." *Science* 349: aac4716.

Sagan, C. 2011. *The Demon-Haunted World: Science as a Candle in the Dark*. Random House.

8 Thinking with Numbers 2: Statistics

There are three kinds of lies: lies, damn lies, and statistics.
—Attributed to Benjamin Disraeli

Do not put your faith in what statistics say until you have carefully considered what they do not say.
—William Whyte Watt, *An American Rhetoric*

Statistics are one of the most powerful tools in biomedical sciences, but also one of the most misused and misunderstood. Statistics allow scientists to distinguish the differences between groups that are due to random variation and chance from those that vary because of a treatment or effect. Estimates vary as to the degree of misuse, but many fields have undergone "crises" related to the number of publications that make statistical errors (Thiese et al. 2015). Statistics are not the sole determinant of whether a hypothesis is true; rather, they are a tool that can lend evidence against or for a hypothesis. To be used effectively, they must be applied correctly.

The Reproducibility Crisis

Although reproducibility is a feature of good science, very often results are published and no one attempts to replicate the results unless they need some value for their own research. Often when replications are attempted it isn't possible to produce the same results as the original research. Replication crises have been found in most fields where large studies have been undertaken to find them. These crises have occurred for a few reasons.

1. Replication studies are typically dull. They aren't cited often and many academic journals will not accept them.
2. They can be time-consuming and expensive, and therefore receive less resources than novel studies.

P-Values

The *p-value* is widely used for significance testing, but is often misunderstood. P-value is used to represent the likelihood that measurements in one group are as different as they are from another group by chance. One group that receives a migraine drug may have on average two fewer headaches per month than one that receives a placebo. The p-value depends on how different the groups are (if the treatment group has ten fewer headaches, that difference is less likely to be due to chance), the number of measurements in each group, and the normality of the distribution of measurements in the groups. Often biomedical scientists will use a p-value of < 0.05 as *statistically significant* (a less than 5 percent chance that the differences between groups were due to chance). Why 5 percent? It is a convention agreed on by many scientists, largely because no one has proposed a better idea.

Statistical significance is not the same as importance, and it is not the same as an effect being real. A statistically significant result is one that is unlikely to be a false positive, but there are circumstances where a statistically significant result can be unimportant, such as when an uninteresting or incorrect question has been asked, or when an incorrect procedure has been used.

Multiple kinds of data testing use p-values. Two of the most common approaches to p-values are Fisher's and Neyman–Pearson's (Perezgonzalez 2015). Ronald A. Fisher believed that the p-value represented a measure of evidence against the null hypothesis (that two groups are not different), so two experiments with $p = 0.49$ and $p = 0.51$ constitute roughly equal evidence that the null hypothesis is not true. To Jerzy Neyman and Egon Pearson, the researcher's task wasn't to have a sliding scale of evidence against the null hypothesis, but rather to make a decision to either accept or reject it. Thus, at the end of an experiment you could either accept or reject the null hypothesis, and you were either accepting a true or false null hypothesis or rejecting a true or false null hypothesis. Thus your goal is to avoid errors termed type I (rejecting a true null hypothesis) or type II (failing to reject a false null hypothesis). In this model you decide the error rate you can live with (call it α) and use p as a comparison to α. If the probability that the results are due to random chance is less than your accepted error rate, you can reject the null hypothesis ($p < \alpha$), and 5 percent is widely accepted as an acceptable error rate.

These two ways of interpreting the p-value often lead to confusion, for example, descriptions of $p = 0.06$ as *nearly significant* or *approaching significance*. In this case the groups don't meet the threshold for significance in the Neyman–Pearson sense, but might provide evidence against the null hypothesis in the Fisher sense. There are circumstances where looking at the data in a Fisher sense makes sense, such as a pilot study where it is not possible to gather sufficient data. In that case the goal is not to accept or reject a hypothesis but to say "this is interesting" or not. Regardless, the researcher should understand which sense the p-value is being used in. Confusion can be avoided if researchers are candid about how they are viewing a statistical test for the purposes of a particular experiment.

In a sense, the seemingly arbitrary $p \leq 0.05$ point of statistical significance is an example of the consensus-building in science, wherein a balance is struck between the risk of accepting a false belief and the risk of rejecting a true belief. However, the p-value is still one mechanism of examining hypotheses among many, and multiple converging lines of evidence are still stronger than a single p-value.

Confidence Intervals

Results are sometimes also reported with *confidence intervals*. Confidence intervals provide more information than p-values in many circumstances. A confidence interval is both a probability and a range for a measurement.

Imagine a trial where it is reported that a drug causes systolic blood pressure to drop in patients in a 95 percent confidence interval of 5 mmHg to 10 mmHg, compared with control patients. This means that there is a 95 percent chance that the true drop in blood pressure caused by this drug falls between 5 mmHg and 10 mmHg. First, this gives us information about the magnitude of an effect. If the drug trial reported only p-values, we might know that the drug was effective, but not have a sense of *how* effective. A drug that causes a 5 mmHg or 60 mmHg drop in systolic blood pressure might pass the same significance test. Because the confidence interval depends on the number of measurements and the variance of the measurements, it can also indicate when more experiments are needed. A p-value can be calculated even when an inadequate number of experiments have been conducted, but a very wide confidence interval would clearly indicate the need for more experiments (du Prel et al. 2009). Second, confidence intervals also indicate the *direction* of an effect size. A medication for pain control could pass a significance test by p-value if it caused *more* pain. Moreover, confidence intervals can indicate statistical significance if the range does not overlap a null effect.

P-values have some advantages over confidence intervals: They're easy to interpret, can be interpreted in a binary fashion, and can help to distinguish a small effect. However, unscrupulous researchers have used p-values to obscure small sample sizes or to argue for the use of treatments that pass significance tests but have very small magnitude of effect such as acupuncture for lower back pain.

Power

Sample size matters to significance tests, and p-values sometimes obscure inappropriately small sample sizes. How do we know what *is* an appropriate sample size? It can be calculated using a "power analysis." This should be done prior to a study being conducted, because it avoids problems that can occur in cases where researchers conduct experiments until they have a statistical test that appears to reject the null hypothesis, and then stop conducting experiments. Appropriate calculation of power prior to conducting a study is rare, and power analyses are often unreported.

Often the actual sample size is dictated by factors other than statistical power. Research funding, the desire to minimize harm to laboratory animals, or a discussion between a principal investigator and a student who wishes to graduate can contribute. While these "externalities" can influence how research is conducted, they shouldn't be the main determinant of how a study is planned out.

The power of a test can be thought of as the probability that the test correctly rejects the null hypothesis, while an alternative hypothesis is also true. The power of a test (and necessary sample size to achieve desired power) depends on the variance of each measurement group being compared, the false negative rate the researcher deems acceptable, and the difference between means (effect size) that the researcher wants to be able to detect.

Power analyses are sometimes exploited with a post hoc application, where researchers conduct the number of experiments they are able to or want to, then back-calculate the parameters of a power analysis that returns that sample size. This methodology can allow researchers to conduct experiments until they achieve the results that they want—which throws the results into question.

The appropriate way to perform power analysis is a priori, before the experiment is conducted. Some scientific journals have been pushing for *prior registration* of power analysis before studies are conducted. This has become more common, especially with large-scale clinical trials.

P-Mining

Another way that p-values have been misused is *p-mining*, data dredging, or *p-hacking*. This happens when researchers test too many hypotheses. The conventional significance criterion of $p < 0.05$ represents a 1 in 20 chance that the two groups being compared are different due to random chance. If you conduct forty such experiments, then you would expect two of those experiments to have $p < 0.05$, even if none of the groups tested were different from the control group. The more hypotheses that are tested, the more false positives you would expect, and if you test hundreds of hypotheses, then you expect many false positive results. This can result in situations where many more false positive results are expected than true positives.

Consider an experiment where RNA microarrays are used to examine the expression of genes in tumor cells compared with non-cancerous tissue. If the experimenter compares expression levels of 20,000 genes, it is expected to find 1,000 false positive "statistically significant" differences. There may only be fifty or so true positives, thus the "signal" is swamped by "noise," or false positives. One solution is to increase the stringency of these tests, requiring much smaller p-values to pass significance testing—which may vastly increase the number of expensive and time-consuming experiments that need to be performed.

It is better to limit the number of experiments done by preselecting a limited number of hypotheses to test based on knowledge of the research area. If only three genes are selected based on biological reasoning, then the number of experiments that need to be conducted to meet a higher stringency to avoid false positives is much lower.

Since many negative results go unreported, the p-value presented to demonstrate significance between two groups may be misinterpreted because of unstated hypotheses that were tested and rejected—making the results more likely to be false positives.

Assumptions

A common error in the reporting of statistics is an absence of reporting assumptions. Many statistical tests assume that two or more measurement groups follow a "normal" distribution. If this isn't true, the test won't produce a trustworthy result. However few publications discuss the distribution of measurements as a factor in selection statistical tests. Other tests rely on assumptions like homoscedasticity (variance that is continuous across a measurement's range), and ignoring this can lead to reporting errors.

Continuous Groups

Often data doesn't separate itself neatly into groups like "treated" and "untreated." A scientist might want to study the relationship between two continuous variables like "blood pressure" and "odds of having a heart attack during a five year period," or "MCAT score" and "medical board scores." There are a variety of approaches to these kinds of studies.

One is to split this kind of data into categories so that significance testing can be used. A researcher might test the null hypothesis "there is no difference in medical board scores between students who scored above a 500 on the MCAT and those who scored below." However, this kind of splitting of categories can create problems. For one, we lose a lot of information about the relationship between the two variables. Second, it allows the researcher to select the point of categorization (why a score of 500+ and not 510+ or 490+?) to find results that are statistically significant.

One way to deal with this is to standardize categorizations used for these comparisons. For example, body mass index

(BMI) has many faults as a tool to understand obesity and body weight in individuals (body shapes and muscle mass differ, for example), but it has utility for studying the aggregate effects of body weight in large populations, and is superior to simply using weight because it does control for one important confounding factor: height. It is common to select any BMI above 25 as "overweight" for such studies, allowing category tests to be used on a continuous variable (do patients who are "overweight" have a higher risk of heart attack compared with patients who are "normal" weight?).

Another (usually better) approach is the use of regressions, which allow researchers to study the relationships between continuous variables. This would allow a researcher to say something like "Only 12 percent of the variation in medical board scores was explained by variation in MCAT scores, but 51 percent of variation in board scores was explained by grades in the preclinical curriculum."

One reason that cutoffs and hard categories are often used in medicine is that they're easy to understand and apply. The statement "A blood pressure of 180/120 mmHg or higher is a hypertensive crisis and requires a hospitalization" is easier to understand and apply than a graph of risk ratios and blood pressures. It is a good shortcut for decision-making, but not for the scientific study of the relationship between blood pressure and risk.

Regressions can also suffer from a kind of problem similar to the multiple comparisons problem. Say we want to arrive at an equation that predicts a student's performance on medical board exams. Among the possible variables, we could include MCAT scores, undergraduate GPA, preclinical curriculum tests, and practice exams scores. Our class only has fifty students, and we use the previous year's student's board scores to develop our model. In the model we include twenty variables, and the equation it produces is quite accurate within our data set, seemingly able to predict board scores within only a few points.

However, when we use the data from the next year's class in that equation, the results are wildly inaccurate. What happened? We *over-fitted* the data. Our function fit one set of data very well, but was useless for any other data set. If we had included fewer variables we might have had a better model!

Deliberately Misleading Statistics

Most of the statistical issues we have discussed are problems that have arisen due to mistakes in thinking about statistics. However, journal editors and readers of scientific papers also have to look out for deliberately misleading statistics. Statistical literacy is low, and this fact can be exploited when work isn't checked.

As we read scientific papers it is important to be appropriately skeptical of the results and methods used. We can't assume that someone describing a statistical test we may never have heard of knows what they're talking about. We need to always check that the appropriate assumptions of each test make sense and have been met.

Simpson's Paradox

Paradoxically, Simpson's paradox is not a paradox. Rather, it is an effect observed in statistics that occurs when an apparent relationship in data reverses when multiple groups are combined.

For example, consider a new drug being tested to treat kidney stones. When the data is first analyzed it appears that the drug has a negative correlation with disease severity. Data was also collected on the size of the kidney stones. When the data is grouped by severity into "small stone" and "large stone" groups, in both groups, the drug is more effective than the older treatment. How is that possible? The drug was tested most often in cases where the stones were larger and less likely to respond to treatment at all.

A data set may show people who are college-educated earning less than those with no college degree. However, when grouped by age, in each age bracket the educated group earns more. If the cohort included a large number of people who just graduated and started their careers, then when the older and younger groups are combined, it can create the illusion of those who are college-educated earning less.

Since Simpson's paradox is really just caused by a failure to account for confounding variables, the term "paradox" is a misnomer. The best way to avoid being fooled by this paradox is to thoroughly consider possible confounding factors when designing studies.

Summary

- Many studies are not reproducible or have not been reproduced.
- Many studies have not calculated needed statistical power.
- P-values and significance testing can be misleading, leading to many "statistically significant" effects that aren't real.
- Many graphs can be confusing or can be used to hide problems with data.
- Often scientists won't make raw data available, even if required to.
- Publication biases mean that often resources are wasted, and negative results are hidden.
- Statistics are sometimes used to obscure negative results.

Exercises

1. Select a medical research paper, find the statistics section, and look up each statistical test. For each test ask yourself:

   a. Was this the appropriate test for this kind of data?

   b. What are the assumptions of this test?

   c. Is there another test that could have been done with this data that might have been more appropriate?

2. Propose a study design. Select a topic that might make for interesting research and then think about how you would approach the statistical methods of the study. What tests would you run? What are some potential confounding factors? What are some limitations? What results would you consider to be "statistically significant"?

Works Cited

du Prel, J.-B., G. Hommel, B. Röhrig, and M. Blettner. 2009. "Confidence Interval or P-Value? Part 4 of a Series on Evaluation of Scientific Publications." *Deutsches Ärzteblatt International* 106: 335–339.

Perezgonzalez, J. D. 2015. "Fisher, Neyman–Pearson or NHST? A Tutorial for Teaching Data Testing." *Frontiers in Psychology* 6: 223.

Thiese, M. S., Z. C. Arnold, and S. D. Walker. 2015. "The Misuse and Abuse of Statistics in Biomedical Research." *Biochemia Medica* 25: 5–11.

9 Science/Conscience: Ethics in Science and Medicine

The saddest aspect of life right now is that science gathers knowledge faster than society gathers wisdom.
—Isaac Asimov

Within every scientist there are two natures: Prometheus and Epimetheus. Prometheus and Epimetheus are brothers. Prometheus had forethought, and Epimetheus only afterthought. Prometheus stole fire from the gods and gifted it to mankind, forever giving humanity the gift of knowledge, toolmaking, and the use of reason. Epimetheus played a role in unleashing all of the evils of the world. Every scientist brings gifts of knowledge to humanity, but also lacks the foresight to control how that knowledge might be used, and what might come of it.

Science has produced fission reactors and fission bombs; penicillin and horrifying human experimentation. Future developments have similar dual potentials. Artificial intelligence may boost productivity or curtail civil liberties, sow discord with false videos, or allow students to submit AI-written essays without having done the schoolwork.

Many attempts to teach ethics in research or medical settings have a blinkered focus on only *professional ethics*. Professional ethics are the norms and behavioral standards defined and enforced by people in the same line of work and by their governing bodies. These are often codified, and these norms can carry negative consequences if violated. Ethics as a field, however, is much broader, and needs to be considered.

Ethics is a field of inquiry, a domain of knowledge, and a set of approaches to problem-solving regarding questions about what is right. Science and medicine present many situations where "the right" is not immediately clear, but some decision must be made.

Descriptive ethics studies what moral beliefs people hold, and *metaethics* studies the nature of morality itself, but we are concerned with *normative* (or applied) *ethics*, which asks, "What is the right thing to do?" There are three common important perspectives.

Virtue ethics defines ethics in terms of virtues. A physician who shows compassion is thought to behave ethically because compassion is seen as a virtue. A scientist who keeps good notes and records is seen as ethically virtuous, because fastidiousness is seen as a virtue.

Using a list of virtues to define ethical behavior has benefits in medicine because it does not require the production of an endless set of rules to fit every application. Imagine a surgeon who spent twenty-four hours in surgery, and despite their best efforts was unable to save the life of a patient. Someone following a list of behavioral rules like "When a patient is bleeding from artery Y and artery X, the bleed in artery Y must be addressed first" must produce an ever-expanding set of rules to address each possible decision. Virtue ethics enables an analysis that isn't as dependent on individual rules. Did the surgeon behave with compassion and do everything that could be done?

There are challenges in using virtue ethics as a sole ethical system. Virtue ethics struggles to provide guidance on which action is good. It frames ethics as a state of being, rather than in terms of actions. If a patient brings a malpractice suit against a surgeon for amputating the wrong limb, the response of "I'm a good person who meant well" is not an adequate defense. When someone is confronted with a person considering termination of a pregnancy, or a decision as to which patient will receive a donated organ, or whether they should allow someone who is suffering without likelihood of recovery access to a lethal dose of morphine, they may find themselves asking, "How would the kind of person I want to be act?" rather than "What is the right way to act?" or "Which actions will produce the best results?"

Deontological ethics considers whether moral actions are "right." Certain actions are always right, and certain actions are "wrong," regardless of consequences. A moral duty can be simple, like "Don't kill," or may be more complicated, like "Don't kill anyone unless it is necessary to protect yourself, your family, or innocent strangers." Deontology allows us to define ideas like "rights," "permission," and "consent," and to develop sets of rules for moral behavior, such as "Do no harm" or "You must have consent for treatment."

Deontological thinking may ignore consequences. A soldier might have a moral duty to follow orders. However, this can lead to the "Nuremberg defense," that a soldier was "*just* following orders," while committing atrocities. A deontological perspective permits ignoring the consequences so long as the action itself has been deemed right. Another weakness of deontological approaches is that they permit effectively arbitrary sets of rules or norms to be established as "right," and then enforced. Both physicians and scientists are obliged to follow certain professional "ethical codes," "oaths," and rules of conduct. However, merely following these rules will not always result in the best outcome. Moral rules developed by others may not be a perfect fit to a given situation, may be rigid and require actions that are cruel, or may prohibit actions that would alleviate suffering.

Consequentialism and *utilitarianism* describe approaches to ethics that involve consideration of consequences and (to a utilitarian) determining the action that will result in the most good (or least suffering). Actions are not always right or wrong, and a person is not a "good person" for possessing virtues. Rather, we consider whether what a person has done has produced good. To a consequentialist, someone who harbors deeply bigoted thoughts but never acts on them and volunteers at a charity does more good than someone who holds no bigoted thoughts but takes no action to improve the world.

However, consequentialism can lead to actions that violate our moral intuitions. A famous example asks if it would be right for a surgeon to end one person's life in order to save the lives of five others with that person's organs. It might be argued that the death of the healthy person and continued life of the sick people was the "most good." However, we don't believe that one person should be killed so that others may live.

Most people use a variety of models to determine what behavior is ethical. In our professional lives we often have deontological rules to follow, from our employers, funding agencies, internal review boards, or religions. Likewise, we seek to exhibit virtues like punctuality and generosity, and to perform consequentialist moral reasoning with regard to the distribution of resources and how patient care should be provided.

A board examining a proposed research protocol involving rats might discuss the proposed work from multiple ethical frameworks. Three moral duties or principles have been widely accepted as pertaining to animal research and were first published by William Russell and Rex Burch in *The Principles of Humane Experimental Technique* in 1959. The three Rs—replacement, reduction, and refinement—are duties that scientists working with animals must adopt. Russell and Burch wanted to change animal research, which they believed caused distress, suffering, or broadly unpleasant mental states to lab animals. Replacement means substituting non-animal substances for animals when possible; reduction means reducing the number of animals to the minimum needed; and refinement means designing experiments to minimize the suffering of animals (Tannenbaum and Bennett 2015). These duties provide a framework for ethical reasoning. The board members can examine a protocol through the lens of replacement, reduction, and refinement.

A utilitarian school of thought concerning animal research comes from the idea that harm caused to organisms capable of suffering must be considered as a part of a cost–benefit analysis compared with the potential good done by the research (Singer 1975). Therefore, the Institutional Review Board (IRB) committee will consider whether the potential benefits of the proposed research will outweigh the harms that the study will do to the research animals. Virtue ethics plays into discussions about the goals and character of the researcher (Walker 2021).

Similar discussions play out in medicine when discussing topics such as the allocation of resources like donor kidneys. Those who smoke cannabis are often ineligible for transplant, while those who smoke tobacco and drink alcohol are not (Minelli and Liang 2011). Thus, transplant facilities make judgments about patient virtue. Likewise, allocation decisions take into consideration patient age and the number of years a donor kidney might be in use, a consequentialist viewpoint about the most good. They also follow strict rules set up internally to make these decisions.

One of the biggest influences on medical ethics was one of its greatest failures: the Tuskegee Syphilis Study, a study conducted between 1932 and 1972 on 399 African American men with syphilis, and 201 without. Nominally, the purpose of the study was to learn about the effects of untreated syphilis. The men in the study were misled about their diagnosis, and none were treated for syphilis, despite the discovery of antibiotics effective for its treatment during the course of the study. During the study, over 100 men died from syphilis or its complications, and they infected others. When the study became publicly known and was terminated, it became clear that formal guidelines were needed for what constituted ethical behavior in medical research involving human subjects.

The 1979 Belmont Report sought to identify ethical principles that would protect human subjects of research. These principles/duties included:

- Respect for persons. This requires that researchers and medical practitioners recognize that every human being has a right to autonomy, and to protect those who cannot make autonomous decisions on their own.
- Beneficence. This idea requires that researchers and healthcare practitioners act in the best interest of study practitioners and patients. This is often summarized as "first do no harm," with the corollaries, "prevent harm," "remove harm," and "practice good."
- Justice. This idea requires that risks, costs, and benefits be distributed fairly. (Miracle 2016)

The Belmont Report informed the rules regarding ethical behavior in research in several US government departments, as well as many professional societies and ethical guides. However, these principles don't cover every situation in which an ethical decision must be made, so there is still an important role for independent ethical reasoning for healthcare providers and practitioners.

Case Studies and Dilemmas

A case study examines a situation in which the ethical course of action is not immediately clear. A dilemma presents a situation where there are two possible courses of action, both of which violate some ethical principle.

Case Study 1

Dr. Green develops a new Alzheimer's drug that shows extremely promising results for neuroregeneration and regaining of memories; however, the patients go through a process of horrific dreams of past memories, especially traumatic memories. Is it permissible to bring the drug to market? Do the potential benefits of the drugs outweigh the harms? Is it ethical to administer the drug to patients who may not be able to consent?

Case Study 2

Dr. Stone is a psychologist conducting research on childhood abuse. Data is collected from many participants, some of which implicates family members in crimes. Each participant has requested anonymity and that their data be kept confidential. Should Dr. Stone report these crimes to authorities? Does giving priority to the research over the well-being of society implicate Dr. Stone morally? Does Dr. Stone have an ethical obligation to intervene?

Case Study 3

During a global pandemic, you are hired to distribute a new vaccine that is extremely effective, but in short supply. You must determine who will receive the vaccine first. One group wants to prioritize based on age, preexisting conditions, occupation, and other risk factors. Another group is willing to distribute the vaccine to wealthier countries first, as the money they pay will fund the creation of new vaccines. Is it ethical to prioritize the wealthier countries if it results in more vaccination long-term? How do you weigh the benefits of each approach? Which populations and individuals should be prioritized?

Authorship

Disputes over authorship are common. Authorship of research papers has value because authorship is used to assess productivity, for grants, awards, jobs, faculty positions, and prestige. Disputes can arise when a principal investigator feels someone did not contribute enough to warrant an authorship, when authorship is granted to someone who did not contribute intellectually to a work, when authorship is "traded" with another researcher, or when it does not reflect their true contributions.

Collaborators might disagree because author order differs between fields. To attempt to standardize these roles, the International Committee of Medical Journal Editors (and others) have proposed guidelines for who should be considered an author of a study (ICMJE n.d.). These include contribution to the design of the work or collecting/interpreting data, drafting the work and revising the intellectual content, having final approval of the published version, and agreeing to be held accountable for the work and answering for its integrity.

Social Responsibility

Differing ideas about social responsibility may also cause disputes. Certain moral theories posit that scientists have a duty to do socially responsible science, or that doing science that benefits society is virtuous. Others argue that if we seek particular truths we find morally acceptable, then we bias ourselves and do bad science. Nevertheless, scientists are capable of choosing which questions they study and how they release their results, and are subject to biased interpretation. These viewpoints address different questions: "Should scientists ignore data they find morally inconvenient?" (No.) And "Should scientists aim to do research that benefits society?" (Yes.) Scientists can practice radical honesty, caring about truth, and openness to changing their minds while working on projects they believe will benefit society.

Conflicts Between Research and Religious Beliefs

Medical research often conflicts with deeply held religious beliefs. For example, human embryonic stem cells (hESCs) are important for work in regenerative medicine, but production typically requires destruction of a blastocyst, causing opposition from anti-abortion activists.

The first anti-abortion movement in the United States was not religious, but tied to the professionalization of medicine, with physicians wanting to be the sole providers of abortion (Mohr 1979). Over time, laws changed, and in the late 1960s and 1970s, restrictions loosened due to feminist thought and concerns over birth defects. Early opposition came from Catholic healthcare groups, eventually involving evangelical Christians as well. By the 1990s, blockades of clinics and acts of violence against providers became common (Reagan 2012).

Because anti-abortion rhetoric has focused on claims that "life begins at conception" and that the moral status of an embryo is equal to that of a full human, the use of embryos in the creation of hESCs became targeted in the early 2000s. President George W. Bush restricted federally funded research to a limited number of preexisting cell lines, severely limiting the creation of new lines. Although those restrictions were eventually lifted, this case exemplifies how conflict can occur between religious belief and scientific research.

Refusal of Treatment and Alternative Caregivers

Common ethical issues arise in medicine around decision-making as patients sometimes have impaired cognition or ability to communicate medical decisions about themselves, or make decisions that physicians may disagree with such as refusing blood products. Physicians must balance principles such as acting in the patient's best interest with acting in accordance with the patient's (or proxy's) wishes.

Case Study: Henrietta Lacks

In 1951 Henrietta Lacks, a cancer patient, was being treated at Johns Hopkins. She had a rapidly progressing cervical cancer. Without her or her relatives' knowledge, a sample of her tumor was given to researcher George Gey. Gey used the cells to make one of the first cell lines, HeLa (from Henrietta Lacks) because the cells had the property of continually dividing and not dying in culture. HeLa cells became widely used in scientific research, and led to developments that saved lives, but were also used to produce financial profit.

It wasn't until twenty years after Henrietta Lacks's death that her family learned about the HeLa cell line. No one had told them that the cells had been taken, used in research, or sold. No one had thought that they should share in the profits from the cells or had a right to privacy of the DNA sequence of the cells.

When the story of HeLa cells became widely known, it became emblematic of white scientists using Black people for experimentation without thought to consent. Many questions remained about her treatment. Would a white patient have been given the option to opt out of the research? Should her relatives have been informed sooner?

Although courts have not recognized people's rights to cell lines derived from their tissues, Henrietta Lacks's treatment violates several of the principles of ethics we have discussed in this chapter. Specifically, the respect for persons, and a person's right to informed consent for how they or their tissue might be used in research. It also illustrates the importance of researchers' consideration of the social circumstances of the populations they're working with. Does the research to be conducted with a population benefit that population?

Summary

- Scientific developments have the potential for positive and negative impacts.
- Professional ethics are norms and standards defined by professional governing bodies, and not adhering to them can have negative consequences.
- Ethics is a much broader field, and we can all benefit from ethical reasoning.
- Normative ethics is concerned with what the right thing to do is.
- Three common approaches are virtue ethics, deontology, and consequentialism.
- Virtue ethics defines ethics in terms of possessing virtues, but does not provide string governance for which actions are good.

- Deontological ethics defines moral duties, but may ignore consequences of an action.

- Consequentialism considers the consequences of an action, but it can lead to actions that violate our moral intuitions.

- Depending on the situation, various approaches can be combined.

- The Tuskegee Syphilis Study led to a rethinking of ethics in medicine and research.

- The subsequent 1979 Belmont Report established the principles of respect for persons, beneficence, and justice.

Exercise

Write two to three paragraphs about Henrietta Lacks. Do the lives saved with research done with HeLa cells justify how she and her family were treated? What could researchers have done differently? Use the ethical perspectives described in this chapter to develop your argument.

Works Cited

ICMJE. n.d. "Defining the Role of Authors and Contributors." https://www.icmje.org/recommendations/browse/roles-and-responsibilities/defining-the-role-of-authors-and-contributors.html.

Minelli, E., and B. A. Liang. 2011. "Transplant Candidates and Substance Use: Adopting Rational Health Policy for Resource Allocation." *University of Michigan Journal of Law Reform* 44: 667–698.

Miracle, V. A. 2016. "The Belmont Report: The Triple Crown of Research Ethics." *Dimensions of Critical Care Nursing* 35: 223–228.

Mohr, J. C. 1979. *Abortion in America: The Origins and Evolution of National Policy.* Oxford University Press.

Reagan, L. J. 2012. *Dangerous Pregnancies: Mothers, Disabilities, and Abortion in Modern America.* University of California Press.

Singer, P. 1975. *Animal Liberation: A New Ethics for the Treatment of Animals*. Random House.

Tannenbaum, J., and B. T. Bennett. 2015. "Russell and Burch's 3Rs Then and Now: The Need for Clarity in Definition and Purpose." *Journal of the American Association for Laboratory Animal Science* 54: 120–132.

Walker, R. L. 2021. "Virtue Ethics and Laboratory Animal Research." *Institute for Laboratory Animal Research (ILAR) Journal* 60: 415–423.

10 Dangerous Deviations: Medical and Scientific Misconduct

Integrity is doing the right thing, even when no one is watching.
—Charles Marshall

Most of the errors in thinking described in this book are typically unintended mistakes and are common. Deliberate misconduct occurs rarely, such as when a researcher fakes data, misuses statistics, modifies results, or plagiarizes. Physicians commit misconduct when they prescribe drugs without legitimate reason, fail to meet the state's standard of care, or commit sexual misconduct (FSMB n.d.).

Cases of outright fraud can disrupt scientific research and cast doubt on entire areas of inquiry. For example, evidence of image manipulation was discovered in decades-old Alzheimer's research, casting doubt on leading theories of the disease (Piller 2022). Scientific fraud has also cast widespread doubt on public health interventions, such as when a now former physician committed scientific fraud, creating a false link between vaccination and autism spectrum disorder in the public consciousness.

The National Science Foundation and National Institutes of Health define scientific misconduct as *"fabrication, falsification, or plagiarism"*:

Fabrication: Fabrication occurs when scientific results are "made up" to match the researcher's desired outcome. For example, researchers tasked to investigate whether participants primed with a word like "intelligent" flashed on a screen for 100 milliseconds are more likely to rate highly the intelligence of a fictional detective in a story. The study is run, but the data doesn't show the statistical relationship the researchers expect. The researchers still believe that the hypothesis is true, and are under pressure to produce results, so they fabricate data that does support the hypothesis.

Falsification: Falsification is altering or omitting certain results so that they fit a hypothesis. For example, modifying an image of a Western blot to appear more or less intense, deleting values from a data set because they appear to be "outliers" without conducting an appropriate outlier test, or retroactively selecting which data sets to include in a study so that the desired results are identified.

Plagiarism: Plagiarism is representing someone else's intellectual work as one's own. This can be as brazen as submitting an essay someone else wrote under your own name, or more subtle as in the case of someone who copies another's ideas, close to verbatim—but does not do a good job of paraphrasing (see chapter 14). The consequences for plagiarism differ based on context. A famous politician can have a book ghostwritten for them and be lauded, while a school essay must be written by the student who turns it in. Plagiarism can also encompass self-plagiarism, such as reusing a figure from one publication in another publication, or copying the introduction from one paper to use in another with minimal changes.

Practices that don't rise to the level of misconduct but which are less than ideal might fall under the category of questionable research practices (QRPs). These include things like hypothesizing after results are known, selective reporting of data, p-hacking, and only publishing studies with positive or significant results.

The prevalence of misconduct is difficult to estimate. Relatively few cases of misconduct are investigated (Reynolds 2004); however, many investigators report "serious misbehaviors" (Martinson et al. 2005). It is difficult to arrive at a solid conclusion about how common misconduct and QRPs are since definitions may vary by field, and those who perpetrate misconduct are unlikely to report the misconduct publicly.

Causes of Misconduct

There are multiple theories that have been put forth to explain research misconduct (Martinson et al. 2005), depending on factors that include the perpetrator's personal reasons, the situation that the perpetrator is in (which may permit the misbehavior), and the external influences that cause the person to desire to commit the misbehavior.

There are three stories frequently told about misconduct— the story of "individual impurity," the story of "institutional impropriety," and the story of "structural crisis" (Haven and van Woudenberg 2021)—each of which places the responsibility on different entities. In stories of individual impurity, we consider the individual as the primary actor in misconduct, and consider their motivations. In institutional stories, we consider the role that institutions play in permitting misconduct, turning a blind eye to misconduct, and/or creating incentives for misconduct. In structural stories, the incentives created by the enterprise of science are examined for the pressures it puts on scientists and the rewards it gives for productivity. How people approach the problem of misconduct may indicate their own motivations in looking at misconduct. If you believe that scientists should regulate other scientists internally, then you are likely to examine individual motivations. If you believe that institutions have the biggest role to play in creating unrealistic expectations of productivity and promoting some scientists over others, then you may look to that explanation. If you're critical of the entire research enterprise, then that explanation is appealing.

Some explanations for misconduct are that misbehavers are people who make a choice to misbehave when the utility of misbehavior is higher than the utility of good behavior. In this model the people who commit misconduct are acting rationally, albeit amorally. Another hypothesis is that some people simply will act greedily when it suits them. Another possible explanation is that people are extremely loss averse, and will behave badly when put in a situation when such a loss is possible, such as loss of a job or research funding (Kahneman 2003).

Most likely each of these factors plays a role in some misconduct: rational choices, loss aversion, and moral failures on the part of those who commit misconduct, as well as institutional and systemic pressures that are put on researchers, which create untenable strain that some relieve through misconduct.

Consequences of Misconduct

Research misconduct can have serious consequences to the individuals involved (Enago Academy 2021), to other individuals, to institutions, and to society at large (NAP 2017). Individual consequences can include loss of a job, loss of reputation, loss of editorship, and even jail time. To institutions there are the costs of wasted research dollars, the costs of investigations, and the costs of additional research undertaken to expand on or understand false results. At the society level, research misconduct can result in a loss of public confidence in science, the spread of conspiracy theories, and wasted time on unfruitful investigatory paths. Research based on misconduct can also cause needless harm to animal subjects and human participants.

Examples of Research Misconduct

The following cases serve as illustrations of the kind of misconduct that can occur in research.

Diederik Stapel was a Dutch social scientist who had at least fifty-eight papers retracted due to fabrication of data (Bhattacharjee 2013). Many of Stapel's papers reinforced appealing ideas about society: that a polluted environment increased racist tendencies, or that eating meat caused asocial tendencies. Stapel excelled, winning awards, grants, and positions. Unusually, he would gather data for his graduate students, who would then analyze it. Eventually, a graduate student grew suspicious when Stapel could not produce the raw data for an experiment. Stapel's fraud poisoned ten PhD dissertations written with him as supervisor, and cast doubt on the journals and editors who had reviewed his work.

Andrew Wakefield was a medical doctor who committed research misconduct in a 1998 paper published in the *Lancet*. The paper claimed the discovery of a new kind of "regressive autism" caused by either measles or the measles, mumps, and rubella (MMR) vaccine. The paper caused a media frenzy that resulted in significant drops in vaccination rates, and more than a decade in claimed links between vaccination and autism. Despite many studies with collectively over 1.2 million patients looking for such a link, no such link was ever identified (Taylor et al. 2014).

Elizabeth Holmes was the CEO of Theranos, a health testing company that marketed a device it claimed could perform
multiple blood tests with only a "drop" of blood, much less
than other testing methods. Holmes was lauded as an innova-
tor and made many media appearances. In 2015 it came to
light that Theranos's device didn't actually work. Commercial
blood analyzers were being used to produce the test results
that Theranos claimed were being produced by its own device.

In 2014 an article was published in *Nature* proposing an
alternate method of inducing pluripotency in stem cells,
known as STAP (Obokata et al. 2014). In the proposed method,
certain adult cells could be converted into pluripotency (the
ability to differentiate into many cell types) by exposure to
certain environmental stressors. This would have been a major
breakthrough in stem cell biology, because previous methods of
producing pluripotent stem cells were more difficult. Multiple
research labs worldwide attempted to replicate the results of the
Nature paper. These replication results failed, and an investigation occurred; the researcher, Haruko Obokata, had falsified at
least some of her results. Obokata maintained her innocence of
misconduct, but agreed to withdraw her article. However, as of
2016, she maintained that "STAP was real" (Goodyear 2016).
Her supervisor, Yoshiki Sasai, committed suicide.

The pressure and rewards system in science can create perverse
incentives that reward cheating. In this sense, scientific misconduct is often not unlike cheating in sports where athletes may
feel that everyone else is using performance-enhancing drugs, so
they must also keep up. Professional science is a highly competi-
tive discipline. Grants, promotions, and accolades depend on
productivity. Scientists may feel pressure to commit small acts
of misconduct. When there are no negative consequences, these
small acts may grow to larger acts of misconduct. Others appear

to be simply uninterested in the truth-seeking aspect of science and simply see it as a way to earn money.

Medical misconduct occurs when a healthcare provider acts in a way that causes harm to a patient, or fails to act appropriately. These can include misdiagnosis, surgical errors, giving the wrong medication, substance abuse, or giving unnecessary prescriptions. Unethical prescribing is the most sanctioned misconduct. Sanctions might include suspension of license or deregistration, deregistration being most common in cases of sexual misconduct (Grant et al. 2021).

Because medicine is a profession (see chapter 5 on self-assessment), it is largely allowed to self-regulate. However, this self-regulation can lead to situations where misconduct is allowed to continue so long as it doesn't affect a particular hospital. The case of Christopher Duntsch became famous through a podcast and television series, *Dr. Death*. The AMA Code of Medical Ethics includes Principles of Medical Ethics, which physicians are supposed to follow, including reporting misconduct by other physicians (AMA n.d.). However, this doesn't always happen. Physicians typically report agreeing with standards of medical ethics, but rarely report colleagues that they believe violate standards of medical ethics (Campbell et al. 2007). Duntsch is alleged to have maimed and killed several patients while working at several hospitals. When a hospital became aware of complaints, rather than report him, he would be quietly fired and allowed to seek employment elsewhere.

Many physicians who have lost licenses or failed licensure due to misconduct have transitioned to other careers as healthcare administrators, consultants, salespeople, or faculty at medical schools. Because physicians are placed in a position of trust over the body, life, and mind of another human being, misconduct can be especially egregious, since it can result in death or permanent injury. An important question that arises from physician misconduct is: How can proper behavior be enforced?

Scientific Racism and Skull Measurements

Very often in the history of science, researchers have brought their own biases to their work, and thus produced flawed results. A prominent example of this is "scientific racism" and the attempt to prove that white Europeans were more intelligent than nonwhites.

There are two kinds of hierarchy in biology. The first is taxonomical hierarchy, which studies the evolutionary relationships between organisms and categorizes them based on those relationships. For example, the group "hominids" includes *Homo sapiens*, but also *Homo neaderthalensis* and *Homo floresiensis*. The group "mammals" includes all humans, but also all dogs, all rodents, and all cows. Because the group "mammals" includes smaller groups such as hominids or rodents, it is a higher-order group in the taxonomic hierarchy.

The second kind of hierarchy is organizational hierarchy. For example, molecules are composed of atoms, cells are composed of molecules, tissues are composed of cells, organs are composed of tissues, humans are composed of organs, populations are composed of individuals, and ecosystems are composed of populations. Ecosystem is a "higher order" of organization than molecule.

Biologists frequently see attempts to produce "rank-order" hierarchies. A rank-order hierarchy involves a moral or value judgment. For example, in a feudal system, a king is seen as closest to a supernatural deity; nobles such as dukes and counts are seen as below the king; and peasants are seen as at the bottom of the hierarchy. Attempts to biologically justify rank-order hierarchies have occurred many times.

These have included attempts to justify the caste system in India, the Confucian social hierarchy, and other social systems, such as by seeking justification for rigid gender roles or laws to maintain rigid social hierarchies of race. Because biology doesn't make value judgments, rank-order hierarchies are not biological.

When researchers have used scientific language, institutions, and respect to justify belief in racial hierarchies, this has been termed "scientific racism." Although hardly scientific, these practices continue to this day to use the window dressing of science to justify racial beliefs and policies.

One such early example of scientific racism was craniometry. Craniometrists believed that they could use measurements of skulls to justify rankings of various human groups by intelligence, temperament, or other abilities. They began with the assumption that such differences exist, and then did experiments attempting to demonstrate that assumption to be true.

The evolutionary biologist Stephen Jay Gould in *The Mismeasure of Man* provided a comprehensive analysis of craniometry's flaws, and Gould's work cannot be improved upon here. Several points that Gould made are especially relevant.

Craniometrists assumed a direct relationship between brain shape or size and intelligence. In reality, the relationship between neural anatomy and cognitive ability is complex and does not only depend on size.

Craniometrists also used potentially flawed methods in measuring skull sizes, allowing their unconscious biases to influence their reported results. Gould reanalyzed results from the craniometrist Samuel Morton and found that Morton's methods were flawed. (Although later authors have challenged

Gould's analysis, the conclusions have largely withstood these criticisms.)

Biased measurements in support of a viewpoint are a *kind of* misconduct, even if unconscious. As scientists, we have an obligation to study our own biases and design our experiments to avoid allowing these biases to influence the results of our research.

Preventing Misconduct

The first step in preventing misconduct is education through training and mentoring. Formal programs such as Entering Mentoring often cover these topics and are mandated by funding agencies or institutions. Institutions can establish offices of research integrity and develop guidelines that will help researchers to understand what constitutes misconduct. The peer review process will ultimately weed out bad science, even if it takes a long time, which is one of the reasons that robust peer review at all levels of the scientific enterprise is important. Likewise, measures that increase research transparency and data availability will make it more clear when misconduct occurs and reduce opportunities for it to occur.

Work can also be done to reduce the incentives for misconduct. Systems of tenure, promotion, and hiring that rely on specific metrics create incentives to target those metrics by "gaming the system." A promotion system that overemphasizes publication count over publication quality would incentivize researchers to seek authorships on publications through less-ethical means such as publishing "least-publishable units," requesting authorship on papers to which they had minimal contributions, or publishing in fringe journals.

A culture of competition can also lead to misconduct by incentivizing cheating, sabotage, "scooping," or other antisocial behaviors. Institutions can focus on developing cultures that encourage collaboration rather than competition.

Stable employment can also lead to greater institutional loyalty, and more interest in adhering to institutional rules. The decline of tenure and increased reliance on contingent faculty such as adjuncts or contract-work professors may incentivize less institutional loyalty and less interest in adhering to institutional rules.

A lack of consequences for misconduct may also provide a perverse incentive. It took more than a decade after the publication of his *Lancet* article for Andrew Wakefield to lose his medical license. Some who have been dismissed from one institution for misconduct have been hired by others despite the history of misconduct. David Sabatini is a prominent biologist who was fired by the Howard Hughes Medical Institute, following alleged sexual misconduct. He then was pledged millions from private donors to start a new lab (Wadman 2023). The case of Christopher Duntsch similarly illustrates this, as throughout his career he did not live up to training expectations, and often botched surgeries, but institutional consequences were rare and most often only took the form of job loss.

Conclusion

Scientific misconduct not only fails to advance knowledge, but actively sets back the search for knowledge, and wastes time and money. Scientific misconduct occurs for many reasons, but is often tied to perverse incentives of the scientific career. Medical misconduct occurs when a healthcare provider acts in a way that causes harm to a patient, or fails to act appropriately.

Misconduct occurs for many reasons, both personal and institutional, and preventing misconduct requires both personal and institutional interventions.

Summary

- Scientific misconduct can be classified as fabrication, falsification, or plagiarism.
- Multiple reasons and motivations may cause misconduct.
- There can be significant societal consequences to research misconduct.
- However, individual consequences, for those who engage in such practices, can vary widely.
- Misconduct prevention can occur through specific training, discussion, mentoring, and peer review.
- Build a culture that combats motivations promoting scientific misconduct.

Exercise

Research the case of Andrew Wakefield.

1. Can you identify the misconduct in that work (fabrication, falsification, or plagiarism)?
2. What do you believe was his motivation?
3. What were the individual consequences?
4. Propose a plan how could this have been prevented.
5. What impact did this have on vaccination?

Works Cited

AMA. n.d. "Code of Medical Ethics." https://code-medical-ethics.ama -assn.org.

Bhattacharjee, Y. 2013. "The Mind of a Con Man." *New York Times*, April 26.

Campbell, E. G., S. Regan, R. L. Gruen, et al. 2007. "Professionalism in Medicine: Results of a National Survey of Physicians." *Annals of Internal Medicine* 147: 795–802.

Enago Academy. 2021. "The Effect of Scientific Misconduct on a Researcher's Career." https://www.enago.com/academy/effect-of-scien tific-misconduct-on-researchers-career.

FSMB. n.d. "About Physician Discipline: How State Medical Boards Regulate Physicians After Licensing." https://www.fsmb.org/u.s.-medical -regulatory-trends-and-actions/guide-to-medical-regulation-in-the -united-states/about-physician-discipline.

Goodyear, D. 2016. "The Stem-Cell Scandal." *New Yorker*, February 21.

Grant, N., S. Valentine, J. Majer, and D. M. Taylor. 2021. "Nature and Outcomes of Sanctioned Medical Misconduct in Six International Jurisdictions: A Case Series." *Australian Health Review* 45: 223–229.

Haven, T., and R. van Woudenberg. 2021. "Explanations of Research Misconduct, and How They Hang Together." *Journal for General Philosophy of Science* 52: 543–561.

Kahneman, D. 2003. "Maps of Bounded Rationality: Psychology for Behavioral Economics." *American Economic Review* 93: 1449–1475.

Martinson, B. C., M. S. Anderson, and R. de Vries. 2005. "Scientists Behaving Badly." *Nature* 435: 737–738. http://dx.doi.org/10.1038/43 5737a.

National Academies Press (NAP). 2017. "Fostering Integrity in Research." https://nap.nationalacademies.org/read/21896/chapter/9, doi:10.17226/21896.

Obokata, H., Y. Sasai, H. Niwa, et al. 2014. "Bidirectional Developmental Potential in Reprogrammed Cells with Acquired Pluripotency." *Nature* 505: 676–680.

Piller, C. 2022. "Blots on a Field?" *Science* 377: 358–363.

Reynolds, S. M. 2004. "ORI Findings of Scientific Misconduct in Clinical Trials and Publicly Funded Research, 1992–2002." *Clinical Trials* 1: 509–516.

Taylor, L. E., A. L. Swerdfeger, and G. D. Eslick. 2014. "Vaccines Are Not Associated with Autism: An Evidence-Based Meta-Analysis of Case-Control and Cohort Studies." *Vaccine* 32: 3623–3629.

Wadman, M. 2023. "David Sabatini, Biologist Fired for Sexual Misconduct, Lands Millions from Private Donors to Start New Lab." *Science*, February 3.

11 The Science of Deception: Pseudoscience and "Alternative" Medicine

> One unerring mark of the love of truth is not entertaining any proposition with greater assurance than the proofs it is built upon will warrant.
>
> —John Locke, *An Essay Concerning Human Understanding*

> The fight against pseudoscience is weakened if trusted medical institutions condemn an evidence-free practice in one context and legitimize it in another. We need good science all the time.
>
> —Timothy Caulfield, Chair of Health Law and Policy, University of Alberta

Pseudoscience and *alternative medicine* are pervasive and fallacious approaches to inquiry and medicine that imitate the real deal, but don't meet the demarcation criteria we discussed in chapter 1.

In reality there is no such thing as "alternative medicine" (Louhiala 2010). For a treatment to be seen as "alternative" it must exist in opposition to an orthodoxy, but medicine as it is currently practiced is defined by science, and depends not on orthodoxy but on evidence. If a treatment can be shown to work (better than other treatments), using the rules of evidence and scientific inquiry it will be adopted as a part of medicine. If it cannot, then the appropriate term is "not medicine," pseudomedicine, or *quackery*.

Not everything that isn't medicine is quackery, so others have proposed a broader taxonomy of "not medicine." Additional categories should exist for treatments that meet scientific standards of evidence, but are not accepted as standard of care in medicine for reasons such as being less effective than other treatments, and treatments that are still being evaluated (Barrett 1998). Suppose there is an accepted surgery for back pain that provides relief in 50 percent of patients and worsens the condition in 3 percent. A new surgery is developed that provides relief in 75 percent of patients, and worsens the condition in 1 percent. A surgeon who does not stay updated on the most recent literature and still offers the first surgery is not practicing quackery, even though the new treatment is better.

"Alternative medicine" frames itself as an alternative to "medical orthodoxy." Samuel Hahnemann, founder of the unscientific and at best baseless practice of *homeopathy*, coined the term "allopathy" to deride the prevalent medical system at the time, which would later evolve into current medicine. Hahnemann believed that medicine was too focused on treating symptoms of diseases, rather than root causes. However, rather than treating root causes, Hahnemann treated nothing. The term has remained in use by practitioners of pseudomedicine to refer to actual medicine. In the United States the term is also sometimes used by real physicians to differentiate between physicians who trained at MD-granting institutions ("allopathic") compared with DO-granting institutions ("osteopathic"). At this point that distinction is largely irrelevant since the curricula are largely aligned and physicians with both degrees share residencies; however, DO-granting institutions once followed the fallacious theories of osteopathy founder Andrew Taylor Still, rather than real medicine. Thoughtful people should still avoid the term for its historic connection to quackery.

How Do We Know if a Treatment Is Pseudomedicine or Medicine?

Similar to our desire to separate science from pseudoscience, it is useful to develop tools to distinguish between medicine and pseudomedicine. In one sense this seems straightforward. Because medicine is an applied science, the ideas we've discussed about demarcating science from pseudoscience apply equally to medicine. However, in many ways, medicine is a special case because of the heightened consequences of intellectual error.

A pseudoscientific belief such as astrology may have negative consequences (or positive), but impacts of an untrue belief in medicine are immediate, obvious, and consequential: People can die or suffer debilitating, lifelong injuries as a result of believing things that are not true. This is why, among the goals of this chapter, is developing an understanding of the different ways that ideas can be untrue, working through examples of cases with ambiguous evidence, and analyzing them to determine if they are premised on true beliefs or false beliefs. If premised on false beliefs, we will examine what kinds of false beliefs, and the source of the errors in the beliefs.

To begin our analysis we select *acupuncture*. Acupuncture is popular and physicians can even receive continuing education credit for acupuncture training, and take elective courses in acupuncture during medical school (Jocham et al. 2017). Beyond that acupuncture has other advantages for our analysis. There are studies that seem to support efficacy of acupuncture for certain functions, while others don't. Acupuncture also has a physical mechanism that could potentially have an effect, unlike other alternative medicine practices like reiki, energy healing, or homeopathy.

Let us break down our analysis into several independent questions: Does acupuncture work? If it does not, what kind of untrue thing is it? If it does not, why do people believe that it does?

First, does acupuncture work? This question demands another question be answered first: What does it mean for a medicine to "work"? We expect a medicine that "works" to improve some kind of medical complaint, and not make it worse. We would expect that it does so better than a *placebo*.

The *placebo effect* means that someone who is expecting to be treated will often experience a change in their symptoms, especially pain symptoms, even if the treatment has no mechanism. This does not mean that the effects of the *placebo* aren't real; there can be real physiological changes. Proper controls and sham treatments are important for examining a medical treatment because the placebo effect can improve patient outcomes. A treatment that improved patient outcomes only through the placebo effect cannot be said to "work," any more than sugar pills could be said to "work."

A medical treatment may "work," but not as well as treatments that are already the standard of care. For example, suppose we developed a new device that served the same function as an adhesive bandage at keeping bacteria out of minor wounds. However, also suppose this device cost $100 per application, caused rashes in 5 percent of users, and emitted a high-pitched "YIP" sound in irregular intervals. This treatment could be said to work in one sense, but it would not make sense to use it in place of adhesive bandages.

What does acupuncture claim to treat? To substantively analyze the effectiveness of a treatment, it is necessary to know specifically the effects it is believed to have. A hypothesis must have a testable prediction. Some sources like the University of California at San Diego medical school make a long list, including back pain, hypertension, hypotension, dental pain, arthritis, and strokes (Center for Integrative Medicine n.d.). The website of the National Center for Complementary and Integrative Health (NCCIH) focuses on pain relief, mentioning lower back pain, neck pain, knee pain, and headache. For the sake of brevity, let's focus on lower back pain. This topic has received a fair amount of study, and although other sources make much broader claims, this allows us to formulate a starting point. Our testable hypothesis: Acupuncture is better than placebo at reducing lower back pain. NCCIH lists three reviews in 2008, 2010, and 2012 as pieces of evidence that acupuncture may be effective for lower back pain. The mission of NCCIH is to "define, through rigorous scientific investigation, the usefulness and safety of complementary and integrative health interventions and their roles in improving health and health care." We take this to be a reasonable starting point for investigation, since as an NIH institute the NCCIH is held to a high degree of scientific rigor, and because this is precisely the kind of analysis that we are interested in performing.

1. The 2008 systematic review looked at twenty-three trials with almost 6,500 patients. The review found that "there is moderate evidence that acupuncture is more effective than no treatment, and strong evidence of no significant difference between acupuncture and sham acupuncture for short-term pain relief." This trial agreed that there was a small, clinically insignificant pain relief associated with acupuncture;

thus, it was no better than the sham control. The results were not necessarily better than the effects of "conventional" treatments.

2. The 2010 review examined the outcomes of eight systematic reviews and meta-analyses (Hopton and MacPherson 2010). It found that "In general, effect sizes (standardized mean differences) were found to be relatively small" (compared with sham acupuncture).

3. The 2012 review looked at twenty-nine randomized controlled trials with almost 18,000 patients experiencing different kinds of pain (Vickers et al. 2012). Although the patients receiving non-sham acupuncture experienced less pain, the effect sizes were 0.15–0.23 standard deviations comparing sham acupuncture to acupuncture, and 0.42–0.57 standard deviations compared with no-acupuncture controls. An effect size of 0.2 would mean that the mean of the acupuncture group was 0.2 standard deviations different from the sham group. An effect size of 0.2 is generally considered to be a "small" effect size (Kim 2015), although this is a rule of thumb arrived at through statistical experience rather than a hard and fast rule. Typically, an effect size is interpreted in the context of a predetermined threshold of clinical significance. For example, in a situation where the effect size was small, but the risk of the treatment was also small and the potential reward was large, then such a small effect size might be clinically important. How do we interpret an effect that is statistically significant, but with an effect size that becomes small when a placebo control is used? There are multiple possible interpretations. The first is that acupuncture can relieve pain slightly better than placebo. The second is there is unaccounted systemic experimental error that recurs in many attempts to test the efficacy of acupuncture for back pain.

Both remain possible. A 2009 review found similar results (Madsen et al. 2009).

This is the best evidence offered for acupuncture as a treatment for lower back pain, each with similar findings: a real effect that works about as well as placebo "sham" acupuncture.

Ultimately, these three pieces of evidence provide very little support for acupuncture. They suggest that there may be a small analgesic effect that is difficult to distinguish from experimenter bias due to its small size, but that small effect is also largely presented when someone is faking acupuncture.

How we choose to deal with this ambiguity is ultimately a matter of judgment. We may choose to focus on acupuncture doing better than no acupuncture. Or we may choose to focus on acupuncture doing little or no better than sham acupuncture. However, our stated hypothesis— acupuncture is better than placebo at reducing lower back pain—is at most weakly supported.

Our next question is: Is acupuncture better than the standard of care? Assuming that the small effect size in the studies of acupuncture is due to a real effect, and not small systemic experimental errors, does it have advantages over other forms of pain management? Even if we do not make this assumption, perhaps the placebo effect is superior to other forms of pain management.

This is again a matter of judgment. Certain mechanisms of pain management through pharmaceuticals are addictive or can cause nausea or other, more severe, side effects. Some of these mechanisms may be ineffective for some patients. In these cases, the relatively small risks of infection or a punctured organ from acupuncture can be largely ignored, and we may find no major harm in patients seeking this treatment.

Certainly, there are other treatments backed by ambiguous evidence that are accepted as a part of medicine. For example, selective serotonin reuptake inhibitors (SSRIs) are widely pre-scribed for mental disorders such as major depressive disorder (MDD). SSRIs can carry significant side effects, such as weight gain, sexual dysfunction, or gastrointestinal issues. A 2017 sys-tematic review found that SSRIs had statistically significant effects, but did not meet the predefined standards for clinical significance (Jakobsen et al. 2017). Large studies of SSRIs have often found them to do no better than placebo at treating MDD (Jakubovski et al. 2016). Other studies have suggested little effect in mild to moderate depression, but some success in cases of severe depression. Still other studies have shown more positive results. Nevertheless, the evidence for SSRIs for MDD is ambiguous, if somewhat better than acupuncture for back pain.

Does this mean that because SSRIs are an accepted treatment, with somewhat weak evidence, that medical care should accept other treatments with somewhat weak evidence? Again, this is a matter of judgment. We may judge that SSRIs are weakly evi-denced and choose cognitive behavioral therapy instead when we face depression, or choose to seek other solutions to back pain when it presents itself. Or we may accept both. Or one but not the other.

Critical thinking about these topics often requires judgment about ambiguous evidence, and the ability to rethink conven-tional or unconventional beliefs. As such, what defines us as critical thinkers is not whether we accept or reject particular treatments, but that we take a systematic approach to analyzing them, fairly evaluate evidence, and attempt to eliminate our own biases.

Acupuncture is not the only "alternative medicine" with weak or no evidence. Deep investigation into the evidence for other alternative medicines tends to yield similar or worse results. Practices such as naturopathy, chiropractic, crystal healing, reiki, homeopathy, and many others claim to be able to heal human diseases, but can't offer the science to support those claims.

Recognizing Pseudoscience and Crackpottery

How do we know when ideas are "crackpot" or "baloney"? Carl Sagan provided his "Bologna Detection Kit" in the *Demon Haunted World* (Sagan 1996):

1. Wherever possible there must be independent verification of the facts.
2. Encourage substantive debate on the evidence by knowledgeable proponents of all points of view.
3. Arguments from authority carry little weight (in science there are no "authorities").
4. Spin more than one hypothesis—don't simply run with the first idea that caught your fancy.
5. Try not to get overly attached to a hypothesis just because it's yours, or you happen to like the presenter.
6. Quantify, wherever possible.
7. If there is a chain of argument, every link in the chain must work. Occam's razor—if there are two hypotheses that explain the data equally well choose the simpler.
8. Ask where the hypothesis can, at least in principle, be falsified (shown to be false by some unambiguous test). In other words, is it testable? Can others duplicate the experiment and get the same result?

The Appeal of False Knowledge

Anyone can make a scientific discovery, but in practice the degree of knowledge and skill required to participate in most science has increased significantly over the last century, making it quite difficult for independent scientists to make contributions in certain fields. The world is still ripe for discovery, but science is hard.

Nevertheless, people trust science and want the word "science" to be associated with the things they sell. *Pseudoscience* offers an easy solution. It has the credibility of science without the effort to learn and practice it. Practitioners of *pseudomedicine* don't have to go through the difficult process of medical school or residency. Even though they may have a "doctorate" in homeopathy or chiropractic, the rigor and honesty of scientific medicine is absent.

Pseudomedicine practitioners and pseudoscientists may feel that it is unfair that scientists and physicians dominate knowledge and medicine. However, scientific medicine is effective, while many other practices aren't. Nevertheless, there will always be motivation to take shortcuts.

Science Denial

Belief in pseudoscience and alternative medicine can turn into science denial when it becomes dogmatic and resistant to criticism, and someone denies a psychologically uncomfortable but scientific truth. This takes many forms such as climate change denial, vaccine denial, or denial of the safety of bioengineered foods. Science denial has certain recurring characteristics: fake experts, logical fallacies, impossible expectations, cherry-picking, and conspiracy theories. These rhetorical tools not only are markers of science denial, but can sow confusion when someone is trying to convert others to denialist positions.

1. *Fake experts* are people who might have apparent medical or scientific backgrounds but who hold opinions that go against scientific consensus. Philip Morris used such "experts" to

confuse discussions of the tobacco risks (Diethelm et al. 2005). During the COVID-19 pandemic "America's Frontline Doctors" pedaled unproven treatments and created confusion about public health measures (Bergengruen 2021). Fake experts can make it seem as if there is widespread disagreement on important topics when, in fact, a minority opinion has been amplified.

2. *Logical fallacies* have been covered in chapter 4.

3. *Impossible expectations* involve denialists setting unrealistic research expectations, or *moving goalposts*. The tobacco industry created its own strict standards for epidemiological studies called "Good Epidemiological Practice," which were more stringent than those used by actual epidemiologists and allowed them to dismiss research they found inconvenient (Diethelm and McKee 2009). A vaccine denialist requests an unethical interventional trial (where patients are deliberately infected with a disease) before believing vaccine research, and despite being unnecessary to understand vaccines' impact on disease.

4. *Cherry-picking* is selectively reading only studies that seem to support a favored viewpoint. A deep understanding of a phenomenon requires appraisal of the available literature. Scientific studies can produce different results due to varying methods, samples, or unaccounted factors. Given a p value of 0.05 as a significance criterion, at least one in twenty "statistically significant" scientific effects might be false.

5. *Conspiracy theories* attempt to explain complex phenomena as the machinations of bad actors.

The absence of *justification* for beliefs in pseudoscience means that it is both easier to generate than science and easier to shape to appeal to human desires.

Science can't always offer hope for permanent cures, clear answers, or desired outcomes, which can be disappointing. People turn to pseudoscience and alternative medicine because detaching from the constraints of reality can fulfill those desires. Pseudoscientists may provide easy explanations for strange symptoms, with explanations like "chronic Lyme disease," or offer to teach incredible superpowers like reiki distance healing.

How Widespread Are Pseudoscientific Beliefs?

One of the largest sociological surveys to address American paranormal beliefs, the Baylor Religion Surveys, found that large portions of the American population held paranormal beliefs of some kind such as beliefs in UFOs or in creatures such as the Loch Ness monster. Some scholars have sought to understand the origins of paranormal beliefs from a sociological standpoint.

One explanation is the *marginalization hypothesis*. It is widely believed that paranormal beliefs are a feature of those who are marginalized in society. However, this idea has been largely refuted. Paranormal beliefs and beliefs in alternative medicine are widespread, and may be more normal than the exception. Forty-five percent of Americans believe in ghosts and demons, and 13 percent believe that vampires "definitely exist" or "probably exist" (Ballard 2019). Twenty-five percent of Americans believe that the positions of stars and planets can influence people's lives; 41 percent believe in extrasensory perception; 55 percent believe in "psychic or spiritual healing, or [that] the power of the mind can the heal the body"; while 73 percent of Americans believe in at least one paranormal phenomenon (Gallup 2005). Although none of these beliefs are supported by scientific evidence, they are widespread, and a worldview grounded in a basis in evidence and informed by science is a rare exception.

Why Does It Matter?

Medical Neglect

Someone who uses alternative medicine may do so *instead* of treatments that are known to work. This can delay effective treatments or miss critical treatment windows. For example, cancer patients who seek "alternative" treatments have a 250 percent higher risk of death overall, and up to 568 percent higher risk of death in the case of breast cancer (Johnson et al. 2018). Indirect harm occurs when needed medical treatments are neglected due to alternative treatments.

Direct Harm

Some alternative treatments can cause direct harm and may lack an adverse event reporting scheme, preventing proper risk analysis (Ernst 2020). A true *risk–benefit analysis* requires an understanding of treatment risks. Injury and death have been linked to chiropractic manipulation (Ernst 2010; Schram et al. 2001), while some people have had severe injury after taking the so-called miracle mineral solution (sodium chlorite), which has been falsely claimed to be a cure for various diseases (Burke et al. 2014; Loh and Shafi 2014).

In the United States adverse events related to vaccines are reported to the Vaccine Adverse Events Reporting System (VAERS). This is one of several mechanisms used to study the safety of vaccines post-approval, including the Vaccine Safety Datalink (VSD), which monitors electronic health records, and the Post-Licensure Rapid Immunization Safety Monitoring (PRISM) program, which uses electronic health records and insurance claims. There are also required studies by vaccine manufacturers post-approval to determine if new risks are observed that did not become clear during clinical trials. These many mechanisms are in place to make sure that these treatments are safe.

Without reporting mechanisms, alternative medicine can never be recommended, because even if a benefit were discovered, the risk–benefit ratio would remain unknown.

How Do Pseudoscientific Ideas Survive, Despite the Disadvantage of Being Demonstrably Ineffective and False?

In his book *Thought Reform and the Psychology of Totalism*, Robert Jay Lifton discusses *thought-terminating cliches*, short phrases that identify group members and shape thought by ending cognitive dissonance. Phrases like "Well, that's just your opinion" reduce an argument to an opinion equal to the opinion of any other. "Agree to disagree" terminates debate by refusing to acknowledge unresolved issues. Complex human problems are compressed into easily memorized and expressed phrases, which supersede any analysis.

Phrases like "Well, that's your opinion," "Don't overthink it," or "Don't be so negative" can silence dissenting opinions about alternative medicine and are common responses to skepticism.

Anecdotal evidence is another way false beliefs protect themselves. Most people find personal experiences and personal testimony to be more convincing than layered scientific and statistical arguments. Storytelling has existed for the entire history of human civilization, and science and statistics have matured on a scale that can be measured only in decades. If you have a positive experience with alternative medicine, data analysis will probably not change your mind.

Heterodox Medicine

Often the appeal of pseudomedicine is framed in opposition to mainstream medicine. If you are dissatisfied with how you've felt when treated by physicians, then the idea of treatments outside of the mainstream may be appealing. Thus, ideas that are presented as outside the mainstream may in fact be old, previously mainstream ideas that fell out of favor in medicine due to failing scientific testing or implausibility. For example, there are many "alternative" treatments that can be viewed as forms of vitalism.

Vitalism is a belief that life is animated by supernatural "force," and these more modern approaches use different terminology to get at a similar idea. Qi (life energy), prana (life force), vital force, innate intelligence, universal life energy, patient energy field, and orgone energy all represent similar ideas. Most of these are unseen, unmeasurable, and untestable. Often maladies are attributed to a lack of balance or flow. The "life energies" are "manipulated" by practitioners.

That vitalist philosophies have been invented and reinvented in multiple cultures and time periods are testament to their intuitive appeal, and apparent explanatory power. Gaps in knowledge about the causes of disease can be filled by a single, simple explanation. Even clearly fictitious versions of vitalism, such as "the force" from Star Wars, gain popularity.

These explanations are so common for people to hold that they represent the mainstream way in which medicine has been viewed for most of human history. The materialist viewpoint that disease is caused by the breakdown in function of biochemical and biomechanical processes is the true heterodoxy. Yet in the face of tremendous evidence for the materialist viewpoint, and frequent falsification of hypotheses consistent with the vitalist viewpoint, vitalist philosophies continue to flourish.

The Difference Between Lies and "Bullshit"

Critical thought about different kinds of things that are not true requires an expanded vocabulary. Terms like "lie," "humbug," "falsehood," and "nonsense" all have distinct meanings. One of the most important distinctions about untruths concerns the motivation of the teller of the untruth. The philosopher Harry Frankfurt wrote about these categories of untruth in his essay, and later book, "On Bullshit." Such scatological language rarely makes its way into the academic lexicon, which has largely chosen to see Latin-derived terms as more serious and academic than Germanic-derived ones, following the Norman invasion of 1066. However, in this case, "bullshit" is the appropriate term, and one Frankfurt selected deliberately and with purpose because of its directness, visceral impact, and fidelity to the subject matter. In this sense the decision to use this term there and here is a substantive decision, rather than just a stylistic one.

"On Bullshit" differentiated between different types of false statements, which depend on the state of mind of the person making the statement.

1. They did not know it was false and believed it to be true. They were not lying, but *wrong*.

2. They knew it was false and said it anyway, which is *lying*.

3. They did not know or care if it was true or false, which is *bullshitting*.

Bullshitting, the act of stating an untruth without concern to the truth, might be more common in our day-to-day lives than lying.

James Ladyman proposed using the concept of bullshit to understand pseudoscience. Returning to our discussion of the demarcation problem from chapter 1, pseudoscience is a kind of untrue approximation of science distinct from other kinds such as scientific fraud. Ladyman wrote: "As a first approximation, we may say that pseudoscience is to science fraud what bullshit is to lies" (Ladyman 2013). Pseudoscience and bullshit are different from scientific fraud and lies in that lies or fraud make claims that could potentially be tested and demonstrated to be false. Pseudoscience and bullshit resist this by not making refutable claims. While a politician might bullshit by blustering through an answer during a town hall that did not address the question asked, a pseudoscientist may bluster through an explanation of how a claimed medical treatment works by vague appeals to realignment of energy or vibrations. Ladyman concluded that Frankfurt's concept of bullshit was similar to that of pseudoscience in that rather than having a desire to mislead about how things are, neither really tried to say anything about how things are.

Another philosopher, Victor Moberger, similarly argued that pseudoscience is a special case of bullshit (Moberger 2020). In pseudoscience and in bullshit, logical fallacies occur not just often, but systematically. Moberger argues that pseudoscience is bullshit with scientific pretensions. Rather than being defined in terms of whether the material in question is true (as a bullshitter may occasionally produce truth), the lack of conscientiousness toward truth is what defines bullshit.

The concept can be applied in other areas as well. Moberger argued that pseudo-philosophy was another special case of bullshit, and academics have begun to use the concept of bullshit to analyze journalistic phenomena, such as "fake news." For example, the 2006 BBC documentary *Acupuncture* showed a woman undergoing open heart surgery following a treatment of acupuncture instead of general anesthetic (Mukerji 2018). An investigation showed, however, that while this claim was true, she also received sedatives and *local* anesthetics. Rather than an example of bullshit, this was an example of a lie of omission or at least shoddy journalism. The reporters tried to get the story right, but failed to do basic investigation that would have changed the conclusions. Thus, in the same way that science has bad science, fraudulent science, and pseudoscience/bullshit, journalism must deal with bad journalism, fraudulent journalism, and bullshit journalism.

Edzard Ernst and Nikil Mukerji analyzed the alternative medicine treatment homeopathy through the lens of bullshit, attempting to answer whether homeopathy is a pseudoscience (Mukerji and Ernst 2022). They observed that practitioners of homeopathy will often defend or justify the practice with bullshit. However, this is not a sole criterion for pseudoscience. They found that this and other criteria, such as a lack of progress, deficient methodology, and external incongruity (implausibility of an idea given our current knowledge of the world), were also indicators of pseudoscientific status.

Pseudomedicine believers often rely on bullshit. Believers often reject science when it doesn't validate their beliefs. Tom Harkin, the US senator who pushed for the foundation of the Office of Alternative Medicine, which later became the National Center for Complementary and Integrative Health (NCCIH), complained that the research being conducted under its auspices was not validating alternative medicine: "One of the purposes of this center was to investigate and validate alternative approaches. Quite frankly, I must say publicly that it has fallen short. I think quite frankly that in this center and in the office previously before it, most of its focus has been on disproving things rather than seeking out and approving" (Boyle 2011). This is a view of science not as a means to search for truth but as a tool for justifying prior beliefs. Even the studies funded by the NCCIH, often undertaken by believers, have found little to support the therapies they investigated.

The scientist must understand two things about bullshit to be able to deal with it:

1. Bullshitters do not care about truth.
2. *Brandolini's law* (Edelson et al. 2021; Vosoughi et al. 2018): It will take more effort to spread truth than it took bullshitters to spread falsehood.

Pseudoscience in Advanced Education

Pseudoscience infiltrates medical education, both undergraduate and continuing education. Continuing medical education (CME) in the United States is regulated by the American Council for Continuing Medical Education (ACCME), which requires CME to be based on current science. Nonetheless, courses in "alternative medicine" topics still grant CME credits, with groups like the "American Academy of Medical Acupuncture" receiving accreditation (Michigan Medicine 2017; University of Florida n.d.). This emblematic of the lax attitude toward evidence that many have taken, which has allowed various treatments to enter the medical field without the stringent scientific requirements that most physicians believe medicine to hold to. Undergraduate curricula also offer electives in "integrative" (alternative) medicine, such as qi gong, reiki, and homeopathy (Salzberg 2011). Some schools even have mandated courses. Reputable institutes that offer these courses lend credibility to treatments which have not met the standards of rigor that the heightened life-and-death stakes of medicine require.

Common Errors in Thinking about Pseudomedicine

Edzard Ernst identified several "follies and fallacies" about alternative medicine (Ernst 2013). Above we have discussed cognitive biases (chapter 3) and logical fallacies (chapter 4). The "follies and fallacies" are frequently found in arguments for alternative medicine, but may be found in arguments for regular medicine, or may be absent. They are not sufficient to determine whether a treatment has met evidentiary standards, but are indicative of flawed thinking on the topic. Ernst included the following:

Argument from popularity: The fact that many people use a treatment doesn't make it right.

Argument from antiquity: Practices being old or used for a long time don't make them more accurate.

Risk evaluation: Adopting a treatment requires a risk–benefit analysis; low-risk treatments without benefits can lose to moderate-risk/high-benefit treatments.

Special pleading: Claims that treatments can't be investigated with science or "western science" attempts to exempt them from scrutiny.

Attacks on profit motive: Criticism of the profit motive in medicine and ignoring the profit motive in "alternative medicine" is a distraction tactic.

Misattributed causation: Recovery can be spontaneous or due to conventional treatment, but attributed to pseudomedical interventions because they occurred at the same time.

Placebo effect: A treatment may have an actual physiological effect through the placebo effect, not through its own merits. Work needs to be done to distinguish between these.

What Believers in Alternative Medicine Need to Do to Meet Scientific Standards of Evidence

To meet scientific standards, alternative medicine believers should adopt the following principles and do the following work:

1. Ideas need to meet scientific standards for acceptance, and practitioners need to be open to change with new information.
2. Scientific publishing and peer review are necessary for quality control and self-correction, but are not foolproof.
3. Openness to criticism is necessary for science; rejecting criticism hampers progress (and doesn't make it go away). Where criticism is inappropriate or flawed, the objection should be well-reasoned and specific, not an accusation of bias sans evidence.
4. Proponents of a treatment or method must demonstrate its virtues; a shift in burden of proof is a red flag.
5. Investigators must be self-skeptical to identify flaws and confounding factors, and to avoid self-deception.
6. To be adopted, new methods need to be better than current treatments.

Evidence-Based Versus Science-Based Medicine

Evidence-based medicine (EBM) bases medicine in evidence, while science-based medicine (SBM) bases medicine in science. In EBM clinical trial evidence is often elevated above other kinds of evidence, and thus endless trials of pseudomedical treatments like homeopathy are undertaken despite no backing in basic science. SBM examines the best science available, not just the best evidence. SBM considers the scientific plausibility of a treatment, and relies less on hierarchies of evidence, especially where randomized controlled trials are not practical. Teaching and learning SBM over EBM can help reduce the creep of pseudoscience into education.

Conclusion

Alternative medicine and pseudoscience offer quick and easy pathways to false authority by those who wish to practice healing or be recognized for making discoveries. On rare occasions, these can lead to new pathways of inquiry, at which point these become real medicine and real science. However, there is no "alternative" to medicine. If it works it isn't, and if it doesn't, it isn't anything.

Summary

- People seek alternative treatments for a variety of reasons, mostly related to how they see it aligning with their personal beliefs.
- Pseudoscience allows the impression of scientific authority without actually doing scientific work.
- The intellectual barrier that treatments must pass in order to be preferred is high enough that even when treatments may have small benefits, they may not be large enough to supply the standard of care.
- Treatments must also be considered with a risk–benefit ratio that can only be understood if risk can be assessed.
- It takes more energy to refute pseudoscientific claims than to make them.

Exercises

1. Identify a practice that is widely considered to be "alternative medicine" and write a reflective paragraph answering the following questions.

   a. What is the evidence for and against this practice?

   b. Is this practice better than no treatment?

   c. Is this practice better than placebo?

   d. Is this practice better than the current accepted standard of care?

   e. What motivates people to believe in this practice?

2. Identify a piece of rhetoric (written, video, audio, or otherwise) where the conclusions aren't true, but the person making the argument doesn't care about truth either way and is not lying (i.e., is using bullshit). Argue why the argument is bullshit.

Works Cited

Ballard, J. 2019. "Many Americans Believe Ghosts and Demons Exist." YouGov, October 21. https://today.yougov.com/topics/lifestyle/articles-reports/2019/10/21/paranormal-beliefs-ghosts-demons-poll.

Barrett, S. 1998. "'Alternative' Medicine More Hype than Hope." In *Alternative Medicine and Ethics*, edited by J. M. Humber and R. F. Almeder, 1–42. Humana Press.

Bergengruen, V. 2021. "How 'America's Frontline Doctors' Sold Access to Bogus COVID-19 Treatments—And Left Patients in the Lurch." *Time*, August 26.

Boyle, E. W. 2011. "The Politics of Alternative Medicine at the National Institutes of Health." *Federal History* 3: 16.

Burke, D., B. Zakhary, and E. Pinelis. 2014. "Acute Hemolysis Following an Overdose of Miracle Mineral Solution in a Patient with Normal Glucose-6-Phosphate Dehydrogenase Levels." *Chest* 146: 273A.

Center for Integrative Medicine. n.d. "Acupuncture/Traditional Chinese Medicine." UC San Diego. https://cih.ucsd.edu/medicine/acupuncturetraditional-chinese-medicine.

Diethelm, P., and M. McKee. 2009. "Denialism: What Is It and How Should Scientists Respond?" *European Journal of Public Health* 19: 2–4.

Diethelm, P. A., J.-C. Rielle, and M. McKee. 2005. "The Whole Truth and Nothing but the Truth? The Research That Philip Morris Did Not Want You to See." *Lancet* 366: 86–92.

Edelson, L., M.-K. Nguyen, I. Goldstein, et al. 2021. "Understanding Engagement with U.S. (Mis)information News Sources on Facebook." In *Proceedings of the 21st ACM Internet Measurement Conference*, 444–463. Association for Computing Machinery.

Ernst, E. 2010. "Deaths After Chiropractic: A Review of Published Cases." *International Journal of Clinical Practice* 64: 1162–1165.

Ernst, E. 2013. "Thirteen Follies and Fallacies About Alternative Medicine." *EMBO Reports* 14: 1025–1026.

Ernst, E. 2020. *Don't Believe What You Think: Argument For and Against SCAM*. Andrews UK.

Gallup. 2005. "Three in Four Americans Believe in Paranormal: Little Change from Similar Results in 2001." https://news.gallup.com/poll/16915/three-four-americans-believe-paranormal.aspx.

Hopton, A., and H. MacPherson. 2010. "Acupuncture for Chronic Pain: Is Acupuncture More than an Effective Placebo? A Systematic Review of Pooled Data from Meta-Analyses." *Pain Practice* 10: 94–102.

Jakobsen, J. C., K. K. Katakam, A. Schou, et al. 2017. "Selective Serotonin Reuptake Inhibitors versus Placebo in Patients with Major Depressive Disorder: A Systematic Review with Meta-analysis and Trial Sequential Analysis." *BMC Psychiatry* 17: 58.

Jakubovski, E., A. L. Varigonda, N. Freemantle, M. J. Taylor, and M. H. Bloch. 2016. "Systematic Review and Meta-Analysis: Dose-Response

Relationship of Selective Serotonin Reuptake Inhibitors in Major Depressive Disorder." *American Journal of Psychiatry* 173: 174–183.

Jocham, A., P. O. Berberat, A. Schneider, and K. Linde. 2017. "Why Do Students Engage in Elective Courses on Acupuncture and Homeopathy at Medical School? A Survey." *Complementary Medical Research* 24: 295–301.

Johnson, S. B., H. S. Park, C. P. Gross, and J. B. Yu. 2018. "Use of Alternative Medicine for Cancer and Its Impact on Survival." *Journal of the National Cancer Institute* 110: 58.

Kim, H.-Y. 2015. "Statistical Notes for Clinical Researchers: Effect Size." *Restorative Dentistry & Endodontics* 40: 328–331.

Ladyman, J. 2013. "Toward a Demarcation of Science from Pseudoscience." In *Philosophy of Pseudoscience*, edited by M. Pigliucci and M. Boudry, 45–60. University of Chicago Press.

Loh, J. M. R., and H. Shafi. 2014. "Kikuchi-Fujimoto Disease Presenting After Consumption of 'Miracle Mineral Solution' (Sodium Chlorite)." *BMJ Case Reports* 2014, November 24. doi:10.1136/bcr-2014-205832.

Louhiala, P. 2010. "There Is No Alternative Medicine." *Medical Humanities* 36: 115–117.

Madsen, M. V., P. C. Gøtzsche, and A. Hróbjartsson. 2009. "Acupuncture Treatment for Pain: Systematic Review of Randomised Clinical Trials with Acupuncture, Placebo Acupuncture, and No Acupuncture Groups." *BMJ* 338: a3115.

Michigan Medicine. 2017. "Creating a Space for Wellness: Integrative Health in Primary Care." *Family Medicine*, March 16–17. https://medicine.umich.edu/dept/family-medicine/events/201703/creating-space-wellness-integrative-health-primary-care.

Moberger, V. 2020. "Bullshit, Pseudoscience and Pseudophilosophy." *Theoria* 86: 595–611.

Mukerji, N. 2018. "What Is Fake News?" *Ergo: An Open Access Journal of Philosophy* 5: 923–946. https://doi.org/10.3998/ergo.12405314.0005.035.

Mukerji, N., and E. Ernst. 2022. "Why Homoeopathy Is Pseudoscience." *Synthese* 200: 394.

Sagan, C. 1996. *The Demon-Haunted World: Science as a Candle in the Dark*. Ballantine Books. https://www.schrijner.nl/sites/default/files/the-demon-haunted-world-science-as-a-candle-in-the-dark-carl-sagan-44ea9c0.pdf.

Salzberg, S. 2011. "Why Medical Schools Should Not Teach Integrative Medicine." *Forbes*, April 21.

Schram, D. J., W. Vosik, and D. Cantral. 2001. "Diaphragmatic Paralysis Following Cervical Chiropractic Manipulation: Case Report and Review." *Chest* 119: 638–640.

University of Florida. n.d. "Integrative Medicine Conference and Workshop: Continuing Medical Education." College of Medicine. https://cme.ufl.edu/integrative-medicine-conference-and-workshop.

Vickers, A. J., A. M. Cronin, A. C. Maschino, et al. 2012. "Acupuncture for Chronic Pain: Individual Patient Data Meta-Analysis." *Archives of Internal Medicine* 172: 1444–1453.

Vosoughi, S., D. Roy, and S. Aral. 2018. "The Spread of True and False News Online." *Science* 359: 1146–1151.

Reading maketh a full man; conference a ready man; and writing an exact man.

—Francis Bacon

Scientific writing is different from most writing, both fiction and nonfiction. There are norms in the scientific literature that can create a barrier to engaging with primary literature directly, such as jargon, unfamiliar experiments, and format and presentation.

The format for scientific papers developed from precursors of letters that individual investigators would write one another to report their findings as part of an "invisible college" of investigators who did not work in the same place (Kronick 2001). At the time, letters and books were some of the main ways of learning about scientific developments. However, letter writing was inefficient because it typically is between two individuals at a time, and books can be inefficient because they require a large chunk of work to be completed before a publication can be made.

Advancements too small to be shared as a book and important enough to be widely distributed could be shared at meetings, but then only those in attendance could benefit. By writing experiments in the format of a letter to a scientific society, results could be shared widely, and be put on record.

Like a letter, modern papers have specific anticipated formats. Many scientific societies still produce journals; however, there has been a significant rise in for-profit publishing entities. Scientific journals publish multiple kinds of articles, including reviews, meta-analyses, commentaries, case reports, and original research, all of which will be briefly described below.

1. *Reviews* are peer-reviewed and published in scientific journals, but they are not original research articles. Reviews contextualize the scientific literature on a topic in a digestible manner, weigh scholarly viewpoints, and provide a bibliography for further reading. Reviews are a useful starting point when approaching a new topic or field because they provide background information that might not be easily available. However, authors select which research to include, so biases can be present. An older review may also miss out on recent developments. Some reviews include specified selection criteria regarding which articles are to be included, but this is not universal (Mulrow 1987).

2. In a *meta-analysis* multiple original research papers are analyzed to identify overall trends, especially if some studies of a topic disagree. A meta-analysis averages the results of many studies, so the individual errors in those studies will become less important. However, the outcome of a meta-analysis still allows for errors. A meta-analysis cannot account for systematic errors in how data is processed, how data is gathered, or a field's thinking about a particular problem. The outcome of a meta-analysis is only as good as the papers and assumptions that go into it. A meta-analysis of studies of acupuncture might seem to show positive results, but upon further

investigation, but the authors included transcutaneous electrical nerve stimulation under the umbrella of acupuncture even though it is different. The preface of *Introduction to Meta-Analysis* relates the story of the recommendation that babies sleep on their stomachs. It was found in the 1990s that babies who slept on their backs were much less likely to die of sudden infant death syndrome (SIDS); however, if a meta-analysis of studies had been performed earlier, many of these deaths might have been prevented (Borenstein et al. 2021). Because individual studies are often ambiguous or conflicting, meta-analysis can be a useful tool for identifying trends.

3. *Case reports* typically report a single or small number of medical cases. We now have more robust research methods, but case reports still serve a few important functions (Nissen and Wynn 2014). They can detect new diseases or events; they can put forward new hypotheses; and they can be useful for medical education. They are limited in that they tend to focus on unusual cases, risking availability bias (see chapter 3 on cognitive biases); they cannot be generalized to other cases; and they cannot establish causal relationships.

4. *Commentaries* are sometimes peer-reviewed, and are published in scientific journals, but they are not scientific articles. Occasionally, a commentary may include pieces of original data. A commentary sometimes accompanies a scientific article to give it context. Commentaries may also be used for an academic to present their viewpoint on a topic, citing recent research. Editors will sometimes also publish special letters attached to articles that present an opposing viewpoint, or with a response made to a claim in a paper that is surprising or controversial.

5. *Primary journal articles* consist of research conducted by the authors of the study. A primary journal article will generally have an abstract, an introduction to explain the research question and hypothesis, a description of what the authors did during the study, a description of the methods used, a listing of results, and a concluding discussion, which allows the author(s) to contextualize their results.

Journals are how most original research is presented in the biomedical sciences. Other fields may publish primarily through conference posters or through books. Journals have a topic, selection criteria, and editors. Editors send submitted articles to referees (generally experts in a field), who then read the draft manuscript and usually make suggested revisions. Authors may be asked to revise the manuscript (sometimes requiring new experiments), may find the manuscript has been rejected from that journal, or may be pleased that their paper has been accepted. Most manuscripts require at least some revision prior to publication. Journals used to be exclusively in print but most are now online, either with an online edition or as obligate online journals. There are different metrics of journal quality, all of which are imperfect. One of the most common is "impact factor (IF)," which measures how often papers from the journal are cited in other journal articles (Saha et al. 2003). In general, a journal with an IF > 10 is considered to be excellent, > 3 is considered to be good, and ≤ 3 to be "average."

The Anatomy of a Scientific Paper

Most journal papers have some variation of the following sections.

1. *Abstract:* The abstract of a paper is likely to be the most read part. It summarizes the contents of the manuscript by

identifying the scientific problem being addressed, briefly explaining what kinds of experiments were done, and summarizing the conclusions. The correct use of an abstract is to decide whether the contents of the paper are of sufficient interest to you to finish reading the whole thing. Taking the contents of the abstract as true without reading the article is not a good practice.

2. *Introduction:* The introduction lets the reader understand why the work was conducted, and gives enough background to understand the experiments that were done. The introduction explains why the experiments were done, and the question that the scientists were hoping to address gives important context to an article. It identifies a gap in knowledge and the broad ideas of the approach taken to fill the gap.

3. *Methods:* In an ideal world, the methods section provides sufficient detail that a knowledgeable reader with their own training and equipment could replicate the experimental methodology to see if they get the same results. This includes details such as the brands of equipment used, the grade of reagents used, the kinds of statistical tests selected, and the steps of any non-standard procedures.

4. *Results:* This section is the narrative description of experiments that were done and the results of those experiments. Most papers in the biomedical sciences tie these results to figures that include graphs and images, as well as tables as a means of summarizing and presenting data. Ideally, the results section should not be used to present any conclusions; rather, it should be used to explain the rationale behind each experiment, what data was collected, and whether it meets pre-set statistical testing criteria.

5. *Discussion:* The discussion gives authors a chance to contextualize their work, placing it in the broader field and discussing

limitations of the work and future directions for research. It also gives authors the chance to speculate about how the results might be interpreted in the light of future evidence.

6. *References:* Ideally, every statement of fact in a scientific paper is supported with either a reference or original data. References allow readers to travel back through the literature on a topic to see how work in a field has progressed and to understand what evidence supports assertions and what doesn't.

Reading a Scientific Paper

Reading a scientific paper to extract needed information should be done by balancing efficiency and thoroughness. There are multiple approaches that might be employed that are not simply reading the paper from beginning to end in a single pass. The goal is not just to accept the claims of a paper, but also to analyze them and ask questions: Is this true? How do they know? Is this repeatable? Were the statistics properly calculated? Is this the best model system? Are there proper controls?

You don't have to read papers front to back. Many may not be relevant to your field or don't contain the information you're looking for. The abstract lets readers quickly decide whether the paper is of interest. The next place to look in a paper is either the introduction or the figures. The figures will contain the summary of the data, a good bit about the experimental and statistical approach, and what the authors think the important takeaways are. This mirrors how many scientific papers are written as well: The authors spend the majority of the time working on gathering data and building compelling figures, and the text is one of the last steps. Finally, if the figures have garnered enough interest, or if there are remaining questions or details you need to understand, read the main body of the text. At this point, having read the abstract and understood the figures, the body of the text will connect dots and add details, but it will be far from as insurmountable as if you start with the main body of the text.

Reading a paper in your research field with critical information will require a much more detailed read than reading a paper outside your field simply out of interest, or simply to extract the description of a method. These are three approaches to how to read a paper.

Three pass approach: This approach involves reading a paper three times, each time with increasing scrutiny. At any point in this process a reader may stop.

1. Gain a general understanding of the paper.
2. Understand the paper's content.
3. Reach a more detailed understanding (Keshav 2007).

In the first pass the reader focuses on the title, abstract, introduction, and conclusion, and attempts to identify the research problem being addressed, the contribution of the paper, the accuracy of the conclusions, and some of the context from its field.

The second pass increases detail, with a focus on figures and methods. The goal of this pass is to be able to summarize the main conclusions of the paper and the evidence presented to justify them, but not focus on exact details of methods or calculations.

The third pass allows the reader to focus on details, such as the specific techniques used, the calculations, the exact meaning of figures, and the implications in the discussion.

The figures first approach: This approach prioritizes examination of the actual data presented in a paper by looking first at the figures presented within it. The reader identifies the conclusion presented in each figure and whether it supports the paper's overall conclusions, and then decides if the data shown in the figures is adequate to support that conclusion. Following this, the reader may check methods, results, or discussion if further information is needed to interpret the figures. This is an efficient way to get the most important information from a paper if there isn't time to read the entire paper.

Skimming: Skimming a scientific paper is an important skill. Often people speak derisively about skimming as though it is less valuable than reading, but it is sometimes appropriate. Skimming is inappropriate if you need to do a deeper reading, such as if you are the reviewer of a paper or want to replicate a method. Skimming can allow you to look through more papers to find the information you need. The most basic kind of skimming is simply reading titles, reading abstracts, and then looking through a paper for figures or keywords that might be interesting or useful.

Don't Believe Everything You Read

An important part of reading the literature is deciding what is worthwhile and what isn't.

Predatory Journals and Peer Review

Predatory journals typically have low or no readership, are not indexed by databases like PubMed, and have low or no impact factors. Learning to distinguish between these is a skill that will develop with your knowledge of a field. Many also deploy deceptive titles that look very similar to reputable journals, making distinguishing harder.

Being "peer reviewed" is not a guarantee of quality. In 2013, a *Science* correspondent published "Who's Afraid of Peer Review," describing a case in which a scientific paper with important errors of method and ethics authored by fictional authors from fictional universities was submitted to 304 open-access journals and was accepted by 157, often with only superficial reviewer comments about formatting (Bohannon 2013). Some journals are lax in their acceptance criteria.

Personal Bias

Personal biases can also influence publications. Some fields are small, and authors who are well known and well liked are more likely to have their papers accepted. In some ways this is formalized by journals that preferentially publish from members of a scientific society or institution. Often submissions are blinded in only one direction, meaning that authors are known to reviewers, but reviewers are not known to authors.

Publication Bias

The *file drawer effect*, or publication bias, occurs when negative results are not published. This can lead to repeated work as multiple groups attempt the same experiments, and to biases in the published literature. Some journals explicitly will not publish negative result, and others don't state this explicitly, but are less likely to accept negative results for publication.

Retracted Papers

One rare means of correcting errors in the scientific literature is *retraction*, in which a published article is marked as "retracted," and the journal that published it no longer views it as meeting its standards.

An example of a retraction caused by flawed methods is the retraction of a paper from *Food and Chemical Toxicology*, first published in 2012, that claimed that rats fed genetically modified food and the herbicide glyphosate developed tumors at a faster rate. Further examination revealed that the paper used flawed statistical methods, because the Sprague Dawley rats used typically develop tumors at a high rate, and therefore a higher number of animals needed to be used in the study.

Hoaxes

Other kinds of publishing controversy have been caused by intentional hoaxers. In 1996, physical professor Alan Sokal submitted a fake paper, "Transgressing the Boundaries: Towards a Transformative Hermeneutics of Quantum Gravity," to the journal *Social Text*, where it was published. The paper made fun of the language used in some cultural studies journals at that time. The author's intent was to question the rigor of the peer review process at the journal, although it is not clear that Sokal made the point he intended. The role of reviewers and editors is usually not to guard against hoaxers and liars, but to improve the quality of submitted work.

Replication

Several fields of scientific inquiry have recently had a *replication crisis* declared, including psychology and cancer biology, where many published results cannot be reproduced. This has been explored further in chapter 7.

Conclusion

Learning to read and interpret scientific literature is a key skill set for emerging scientists and physicians. It is necessary not only for keeping up with new developments but for being able to critically evaluate claims made in the scientific literature, and as a stepping stone to writing original research articles.

Works Cited

Bohannon, J. 2013. "Who's Afraid of Peer Review?" *Science* 342: 60–65.

Borenstein, M., L. V. Hedges, J. P. T. Higgins, and H. R. Rothstein. 2021. *Introduction to Meta-Analysis*. John Wiley & Sons.

Keshav, S. 2007. "How to Read a Paper." *Computer Communication Review* 37: 83–84.

Kronick, D. A. 2001. "The Commerce of Letters: Networks and 'Invisible Colleges' in Seventeenth- and Eighteenth-Century Europe." *Library Quarterly: Information, Community, Policy* 71: 28–43.

Mulrow, C. D. 1987. "The Medical Review Article: State of the Science." *Annals of Internal Medicine* 106: 485–488.

Nissen, T., and R. Wynn. 2014. "The Clinical Case Report: A Review of Its Merits and Limitations." *BMC Research Notes* 7: 264.

Saha, S., S. Saint, and D. A. Christakis. 2003. "Impact Factor: A Valid Measure of Journal Quality?" *Journal of the Medical Library Association* 91: 42–46.

13 Hidden Curriculum

> Education is not preparation for life; education is life itself.
> —John Dewey

There are multiple facets to student experiences that can fit under the umbrella concept of the *curriculum*. Some facets are intentional and explicit, such as learning objectives and course schedules. Other learning experiences can be unintentional and implicitly delivered by instructors or institutions, such as school culture or policies toward diversity. Both explicit and implicit learning can be powerful and shape student attitudes, identity, and performance. Here we will briefly introduce elements of a curriculum in higher education with a focus on exploring the *hidden curriculum*, its potential impact on students, and why, for students, it might be necessary to uncover the hidden curriculum.

Why Is Identifying the Hidden Curriculum Important?

In science and medicine, disparities can be found based on socioeconomic background, personal identity, and race. Women and underrepresented minorities feel less supported during graduate training and find a more difficult hiring environment (Clauset et al. 2015; Stockard et al. 2021). Despite efforts to increase diversity, Black and Hispanic students remain underrepresented in medical school (Lett et al. 2019). Faculty with a doctorate degree (PhD) are more likely to have a parent with a doctorate (close to 22 percent), while most have a parent with a graduate degree (> 50 percent) (Morgan et al. 2022). Medical students are more likely to have a parent with a graduate degree compared with the general population (Grbic et al. 2010). Students from privileged backgrounds (i.e., who have parents with advanced degrees and higher socioeconomic status) have an advantage to accessing and potentially succeeding in graduate school. Academically high-achieving parents can pass along skills on how to navigate elements of the hidden curriculum through their own experiences as well as providing access to networks that are not available to others (Calarco 2020; Morgan et al. 2022). For those individuals who do not come from similarly privileged backgrounds, navigating graduate or medical school and beyond can be more challenging. It is important to recognize aspects of the hidden curriculum so that you can help mitigate some of its influence.

Curriculum and the Hidden Curriculum

Curriculum is a combination of educational goals, objectives, and experiences integrated with evaluation, assessment, and expected and measured outcomes. An institution's or degree program's curriculum is composed of what is formally designed (i.e., courses, learning objectives, educational experiences, and evaluations/assessments), what is taught by faculty in the classroom (lectures or lab sessions), and what is learned by the student. Commonly, degree programs will have curricular maps or sequences published on websites for students to view, especially in highly structured programs such as law or medical school. The formal or intended curriculum includes the goals and objectives of courses and the expected outcomes for students who complete the program. A research-based graduate program may have objectives related to developing critical thinking skills and demonstrating expertise in a specific field of study, with the output being a thesis or dissertation. Courses, evaluations, and assessments would then be intentionally designed to help the student accomplish the program goals and outputs.

There may be a disconnect between what is intended by the institution or program and what is actually taught to or experienced by the student. There is a great deal of knowledge that is transferred or imputed to students that is not explicitly intended through formal curriculum design. This implicit knowledge is commonly referred to as the hidden curriculum (Hafferty and Gaufberg 2013). The content of the hidden curriculum can vary greatly between fields or programs of study but commonly includes social and professional expectations. Students may not be aware of these hidden norms, which can have significant consequences on success. The hidden curriculum can be found in all areas of study, including research, medicine, engineering, and geosciences (Fryer-Edwards 2002; Pensky et al. 2021; Villanueva et al. 2018).

The hidden curriculum, by definition, is not overtly observable. Intentional reflection and purposeful analysis of programs are needed to try to identify the hidden curriculum, but it is ever present. In many ways, the hidden curriculum is the area of education that is "taken for granted" (Calarco 2020). In graduate or professional degree programs, some of the things that may be taken for granted by faculty or mentors are understanding how to navigate research, how to read literature critically, how to write and publish appropriately, professional behavior, and even ethical practice. For experts, these behaviors have become *operationalized*, meaning experts perform the tasks listed without thinking because they have become second nature. Faculty and research mentors may be unaware that their students are novices and need guidance.

Socioeconomic and cultural influences will also be part of the hidden curriculum, with both negative and positive outcomes. Below we describe three scenarios to show how the hidden curriculum can influence cultural or socioeconomic experiences for students for good or ill.

Scenario 1: Medical school is typically split between preclinical and clinical education, with the former devoted to intense instruction in the biomedical sciences. The institution, PhD faculty, and outside licensing exams all highlight the importance of a strong foundation in the biomedical sciences. When students in their preclinical years interact with some of the more senior students or clinical faculty in a clinical setting, they are told that much of the science information they learned is not as important and may not help them become "good doctors." This theme clearly undermines the intended preclinical curriculum and contributes to a negative educational culture.

Scenario 2: A graduate student is excited about attending a conference to share their latest research findings and begin networking with others in their field. They apply for and receive a $250 travel award from the conference organizers. The student calculates it will cost over $1,000 to attend the conference, including registration fees, travel, and hotel expenses. The student's mentor expects the student to attend the conference, while the department assumes the student has the economic flexibility to afford the extra expense as an investment in their professional future. The student is caught between choosing to sacrifice present expenses such as rent or food to attend a conference that can help with future professional growth.

Scenario 3: A medical school creates a mission to improve medical care in underserved communities and train students to serve in areas of need. The school is located within a low-income community and provides state-of-the-art medical education to its students. All clinical experiences and rotations for students are in the local community or in communities with similar socioeconomic status. The school establishes strong ties with local community leaders and creates medical outreach programs providing volunteer opportunities for medical students. The school supports student-created organizations that target underserved communities and promotes students from underrepresented backgrounds. The medical school employs faculty and staff from diverse backgrounds who choose to participate with students in outreach and volunteer activities. The culture of such a school will most likely foster advocacy for and service in medically underserved communities. Students will receive a broad education in the socioeconomic challenges of the surrounding and similar communities. The school understands the power of the hidden curriculum and uses it to their advantage to help educate students and promote the institutional mission.

Taking on the Hidden Curriculum

Success in graduate or medical school can appear to be passed on from one or both parents. It is therefore important for students who do not come from families with advanced degrees, regardless of how supportive they may be, to find resources and build networks. One resource to uncover the hidden curriculum is the contents of this book. The various chapters, from cognitive bias to self-assessment to science writing to study strategies, address many aspects of the hidden curriculum encountered in graduate and medical school. Reading the chapters and working through the exercises should allow for reflection on past experiences and hopefully provide insight for future scenarios. Another resource is *A Field Guide to Grad School*, by Jessica McCrory Calarco (2020). This book also attempts to directly uncover the hidden curriculum most likely experienced by graduate students. Aside from these texts that can lay down a framework of the hidden curriculum, other direct actions can also be taken by students. Finding mentors within and outside the institution with a similar background can be beneficial. These mentors can better advise about the challenges and unwritten expectations of the field. Reaching out to national organizations, such as the National Black Graduate Student Association or the Student National Medical Association, will also help to identify mentors and help create a support network. It may also be helpful to identify and recruit allies at your school to help push elements of the hidden curriculum into sight. Allies can be other students, faculty, or administrators. Communicating with graduate student councils or student government may also be a way to address the hidden curriculum. The basic idea is to build a support network of knowledgeable individuals who can both advise and promote change. Giving back is a final and crucial element. Once you have progressed enough in your degree program, provide support to new students to help them navigate the challenges you may have faced with the hidden curriculum.

Practical Tips

The authors of this book have decades of combined experience working with graduate and medical students with diverse racial and socioeconomic backgrounds. From these experiences we have found the need to promote practical tips and suggestions to help students navigate their advanced degree programs.

Tip 1: Ask for help. Asking for help can be difficult and puts you in a place of vulnerability. The willingness to seek help can be strongly influenced by a student's background. Asking for help in higher education has socioeconomic roots. Students with a more privileged background (higher family income, superior high school education) are more likely to ask for help from faculty and their peers, while first-generation and lower-income students tend to practice more isolated and independent strategies (Calarco 2020; Eaton et al. 2020). Both independent and collaborative work strategies are necessary for success in higher education and beyond. However, the lack of exposure to unwritten rules of higher education can make this more challenging. Reluctance to ask for help may also be justified as research has shown implicit bias by university faculty that favors white males in STEM fields (Eaton et al. 2020; Moss-Racusin et al. 2012). Independence is a strong personal trait; however, collaborative and interactive engagement is more generally valued in higher education, including graduate and medical school.

Tip 2: Failure is part of the process. Although no one wants to fail, it happens to everyone. Experiments may not work out, your hypothesis was off, or you bombed an exam. These should be seen as opportunities for learning. However, learning from failure appears to be an uncommon trait, especially in

Western cultures (Eskreis-Winkler and Fishbach 2019). Negative feedback from failure can cause a person to shut down and be closed off to information that can be corrective. This may be due to "ego." If you can separate yourself and your ego, you will be more likely to accept failure and use it as a powerful learning tool.

Tip 3: It's okay to be wrong and take risks. We all prefer the feeling of being right to help avoid failure. As a novice in a field of study there will be a great deal of knowledge that has not been acquired. There is a limit to one's knowledge, and realizing and accepting this notion is referred to as *epistemic humility*. Science has not necessarily adopted such a philosophical approach and tries to avoid being wrong, which can be referred to as *epistemic risk* (Parascandola 2010). It has been argued that the risk of being wrong in scientific research results in more conservative experimental approaches and reluctance to gather new data for fear of supporting a wrong conclusion (Parascandola 2010). Risk avoidance may also contribute to overdiagnosis in medicine, which can result in significant financial costs along with unneeded treatments and potential harm to patients. There is no hard and fast rule regarding when to take on risk or what is acceptable, but it is something that should not always be avoided.

Tip 4: Develop a professional network. Networking doesn't have to be schmoozing, or approaching strangers in your field to make an introduction out of the blue. Networking is the natural process of developing professional relationships with others in your field who are aware of and interested in your career. You might need a letter of recommendation, someone to send a grant application for review, or someone on a professional society committee who is aware of your work when reviewing an award application. In medicine, cohesive social networks can increase the effectiveness of care (Cunningham et al. 2012). Within scientific research, different network strategies can yield different results. Exploitation strategies develop and use existing networks and can improve research quantity (such as journal articles and grant submissions). Exploration strategies pursue new knowledge and innovation, such as by seeking new professional relationships, which can optimize research quality by incorporating new ideas (Siciliano et al. 2018).

Tip 5: Develop your work–life balance. Life doesn't start after graduate school; it is happening now. Graduate school can take a toll on family life (Gilbert 1982). When considering graduate programs, take into consideration time, money, effects on relationships (Leonard et al. 2005), residency status, characteristics of the academic environment, spousal considerations, and social environment (Kallio 1995). Pursuit of doctoral degrees includes costs to relationships, finances, mental and physical health, and advancement in other careers (Cefaratti et al. 2007). Many doctoral students don't understand the difficulty and potential costs of doctoral programs before entering (Golde and Dore 2001). Talk to people already in the program before joining.

Tip 6: Pick good mentors. The scale of hidden curriculum knowledge ranges from how to address faculty by email to the nuances of navigating interdepartmental university politics. This knowledge is often developed as a part of mentoring relationships, by observing near-peer students, and through trial and error. A good mentor can correct your misunderstandings, explain academic conventions, point out opportunities, and make new opportunities available. A good mentor will also help you monitor your academic progress and make corrections when needed.

Conclusion

The hidden curriculum is a challenging concept to identify but still critical for success in higher education. Here we have discussed approaches to mitigate these challenges and brought to light some examples and solutions to address this. From identifying mentors to networking with people, asking for help, and embracing failure and humility, strategies like these allow a novice learner to navigate the intricacies of the educational world.

Summary

- The hidden curriculum is the implicit information students retain from education.
- Unwritten social, cultural, and professional norms are also part of the hidden curriculum.
- Tips to expose and work with the hidden curriculum include finding mentors with similar backgrounds, learning to collaborate, asking for help, learning to fail, and being willing to take risks.

Exercises

1. Reflect on previous academic experiences, classes, or programs and write down what you learned that was not explicitly taught to you. Share this exercise with others and discuss your answers together.
2. Using your reflection on the hidden curriculum, write down how it has affected you in both negative and positive ways.

Works Cited

Calarco, J. M. 2020. *A Field Guide to Grad School: Uncovering the Hidden Curriculum.* Princeton University Press.

Cefaratti, M. A., C. A. Horkey, B. Flora, et al. 2007. "Opportunity Costs of Graduate Education: An Exploratory Study." *Journal of Continuing Higher Education* 55: 14–23.

Clauset, A., S. Arbesman, and D. B. Larremore. 2015. "Systematic Inequality and Hierarchy in Faculty Hiring Networks." *Science Advances* 1: e1400005.

Cunningham, F. C., G. Ranmuthugala, J. Plumb, et al. 2012. "Health Professional Networks as a Vector for Improving Healthcare Quality and Safety: A Systematic Review." *BMJ Quality and Safety* 21: 239–249.

Eaton, A. A., J. F. Saunders, R. K. Jacobson, and K. West. 2020. "How Gender and Race Stereotypes Impact the Advancement of Scholars in STEM: Professors' Biased Evaluations of Physics and Biology Post-Doctoral Candidates." *Sex Roles* 82: 127–141.

Eskreis-Winkler, L., and A. Fishbach. 2019. "Not Learning from Failure—The Greatest Failure of All." *Psychological Science* 30: 1733–1744.

Fryer-Edwards, K. 2002. "Addressing the Hidden Curriculum in Scientific Research." *American Journal of Bioethics* 2: 58–59.

Gilbert, M. G. 1982. "The Impact of Graduate School on the Family: A Systems View." *Journal of College Student Personnel* 23: 128–135.

Golde, C. M., and T. M. Dore. 2001. *At Cross Purposes: What the Experiences of Today's Doctoral Students Reveal about Doctoral Education.* Pew.

Grbic, D., G. Garrison, and P. Jolly. 2010. "Diversity of U.S. Medical School Students by Parental Education." *AAMC Analysis in Brief* 9, no. 2.

Hafferty, F. W., and E. Gaufberg. 2013. "The Hidden Curriculum." In *A Practical Guide for Medical Teachers*, edited by J. A. Dent and R. M. Harden. Churchill Livingstone.

Kallio, R. E. 1995. "Factors Influencing the College Choice Decisions of Graduate Students." *Research in Higher Education* 36: 109–124.

Leonard, D., R. Becker, and K. Coate. 2005. "To Prove Myself at the Highest Level: The Benefits of Doctoral Study." *Higher Education Research & Development* 24: 135–149.

Lett, E., H. M. Murdock, W. U. Orji, J. Aysola, and R. Sebro. 2019. "Trends in Racial/Ethnic Representation Among US Medical Students." *JAMA Network Open* 2: e1910490–e1910490.

Morgan, A. C., N. LaBerge, D. B. Larremore, et al. 2022. "Socioeconomic Roots of Academic Faculty." *Nature Human Behavior* 6: 1625–1633.

Moss-Racusin, C. A., J. F. Dovidio, V. L. Brescoll, M. J. Graham, and J. Handelsman. 2012. "Science Faculty's Subtle Gender Biases Favor Male Students." *Proceedings of the National Academy of Sciences* 109: 16474–16479.

Parascandola, M. 2010. "Epistemic Risk: Empirical Science and the Fear of Being Wrong." *Law, Probability and Risk* 9: 201–214.

Pensky, J., C. Richardson, A. Serrano, et al. 2021. "Disrupt and Demystify the Unwritten Rules of Graduate School." *Nature Geoscience* 14: 538–539.

Siciliano, M. D., E. W. Welch, and M. K. Feeney. 2018. "Network Exploration and Exploitation: Professional Network Churn and Scientific Production." *Social Networks* 52: 167–179.

Stockard, J., C. M. Rohlfing, and G. L. Richmond. 2021. "Equity for Women and Underrepresented Minorities in STEM: Graduate Experiences and Career Plans in Chemistry." *Proceedings of the National Academy of Sciences of the United States of America (PNAS)* 118: e2020508118.

Villanueva, I., M. Di Stefano, L. Gelles, and K. Youmans. 2018. "Hidden Curriculum Awareness: A Comparison of Engineering Faculty, Graduate Students, and Undergraduates." Paper presented at World Education Engineering Forum, 2018, Albuquerque, NM.

14   How to Write Scientifically

> If you don't have a question, you are not doing good science. If readers can't tell what it is, you are not writing good science.
> —Joshua Schimel

Why is it important to learn to write scientifically? Perhaps you have a writing assignment in a lecture course or need to complete a laboratory class manual. Maybe writing is a requirement in the form of a thesis or dissertation for your degree program. Beyond short-term motivations, scientific writing is an incredibly important form of communication. A scientist or a student may come up with the greatest idea, but it could be lost if it is communicated through poor writing. Additionally, writing can be an effective method of learning, as it requires content knowledge and promotes critical thinking (Quitadamo and Kurtz 2007). Here we will build upon concepts presented in chapter 12 on reading scientific literature and focus on some major aspects of how to approach scientific writing.

Before You Begin

Many books and articles have been devoted to the important topic of scientific writing. We encourage the reader to research more in-depth resources and have provided two example references at the end of this chapter (Knisely 2005; Schimel 2012). Rather than replicate the work of others, here we will stress three important writing points regarding storytelling, time, and exemplars.

Scientific writing is often concise and commonly uses technical language that is absent from editorial opinion, unlike fiction, which can rely on descriptive language. Although scientific writing is meant to present and interpret data, it should still be seen as storytelling. The reader should be interested in your topic and enjoy what you have created. Style depends on the purpose of the writing and the target audience.

The way you approach writing a term paper for an undergraduate biology course will differ from a manuscript for publication or a grant proposal. One of the main differences is the time you take to get to your main idea. In a grant proposal you want to hit the reader or reviewer with your big idea at the very beginning, to capture their interest and get them hooked on your hypothesis. You can then provide necessary background information to support the idea or hypothesis. A common format for a grant is first to have a section on significance, followed by background information, possibly preliminary data, and then experimental design. In formats such as manuscripts or books, there can be more time before revealing the main idea or purpose. Sufficient background information is presented first that allows the reader to put the manuscript's purpose into context. Many journal formats fit this outline, with an introduction section followed by the methods, results, and discussion (IMRAD). Within the introduction section, the statement of purpose is commonly found toward the end of the section. Therefore, when planning your "story," keep in mind the audience, the format, and the placement of the thesis or main statement before you begin writing.

Writing a manuscript will take significantly longer than you think. Just ask a classmate who had to write a biology paper or honor's thesis. To write effectively, sufficient time should be given to three processes that need to be conducted repeatedly: writing, review, and revision.

To prepare for the writing step, read as much as you can of articles, books, and reviews about your subject in order to become familiar with the terminology and background information. While reading, start thinking about what to write and how the information you are learning can be used. Next, create an outline. What is it that you want to say and in what order should it be presented? Review the outline several times, make revisions, and provide additional details with every revision. A more detailed outline can make the writing experience easier. Once the foundation has been established with an outline, start writing. Sometimes the most difficult part is starting, but write whatever you can, whether it is good or not. This allows you to start reviewing and revising, which can be easier than starting from scratch and may even help you push through writer's block (Schimel 2012). Once there is a solid draft, ask peers or mentors to review and provide feedback. It can be difficult to share your writing, but appropriate feedback from peer review is a critical component to successful writing. Use suggestions from others to work through the revision process and improve your writing. Through the reiterative process of writing, reviewing, and revising, you can create a written product you can be proud of.

Exemplars (ideal models) can aid the writing process. Every science field has its own accepted format and style for publication. As outlined in chapter 11, it is advisable to become familiar with several of the "go to" journals in your field and to pay attention to the styles and formats of writing. Discuss styles of writing with classmates or colleagues and, most importantly, request feedback from mentors or faculty on your writing. It is appropriate to ask for writing examples from peers or faculty mentors. Examples of papers or grants can be used as writing models to establish writing expectations and to provide inspiration, as well as details regarding style or format. However, examples are just that, examples. Your writing project should be your own thoughts and language and not copied from examples or other sources. This may seem obvious, but the task of appropriate paraphrasing with citation to avoid plagiarism is not always clear to new writers.

Key to Science Writing: Putting It in Your Own Words!

For novice writers of science, one of the biggest challenges in a writing project is trying to put new and complex language into one's own words and style. Often, a novice writer will try to emulate expert language without a mastery of the required knowledge or terminology. This can result in improper paraphrasing or citation, and plagiarism.

The last few years have seen major advances in generative artificial intelligence, in the form of large language models (LLMs). LLMs are artificial neural networks trained on large data sets of human-written text, and are capable of producing text that is relevant to human-provided prompts. This is done by using weights and values stored in the trained network to predict a series of the most probable sequence of words. Because LLMs can generate text, and answer questions based in its training set, it has rapidly become a tool used by many in writing and research projects. However, caution needs to be exercised when using these tools, for several reasons. First, LLMs are capable of "hallucinating," which is when they generate content that is false or inaccurate. The data in a LLM is stored as weights and values in a neural network, rather than directly in a database, so independent verification of fact is necessary. For example, early versions of GPT 3.5 would produce citations that appeared to be real, but were false. The LLM had learned the patterns of what a citation looked like but was not testing veracity. This has caused a number of well-publicized problems, such as legal briefs being filed that cited cases that did not exist. Another issue with the use of LLMs in writing is that if the author is not clear that they are using an LLM in producing the text, they may be guilty of plagiarism as the work they are presenting is not their own.

However, there are a number of scenarios in which a LLM may be used productively and usefully. For example, a LLM may provide help in outlining, breaking writer's block by suggesting ideas, and providing constructive criticism of text and suggestions for improvement.

Several definitions of plagiarism have been developed, and all are variations of a common theme (excellently reviewed in Nelms 2012 and Roig 2015). Plagiarism can be defined as the use of words, ideas, images, or material and claiming them as one's own without acknowledgment of the originating source. Where definitions of plagiarism may differ is in context and intent. The Council of Writing Program Administrators (2019) defines plagiarism as "deliberate use," implying there must be intent by the plagiarist. However, it has been argued that intent should not be a characteristic of plagiarism and the focus should simply be on ethical practices of writing and training in paraphrasing (Nelms 2012; Roig 2015). For our purposes, we will not be arguing the finer points of what is and isn't plagiarism but will provide examples of types of plagiarism and how a new writer to science can work to avoid them.

Broadly, plagiarism can be categorized as either intentional or unintentional. Intentional plagiarism is the purposeful copying of text or use of ideas without attribution of the original source. Several reasons have been proposed why students would engage in such unethical writing, including pressure for a good grade, not giving sufficient time to the writing process, or a lack of motivation to appropriately paraphrase and cite (Nelms 2012). Unintentional plagiarism is when original material is copied or not sufficiently changed by a student. As with intentional plagiarism, the reasons for unintentional plagiarism are many. They include lack of knowledge of when and how to cite, lack of content knowledge or terminology, poor note taking that doesn't put information into one's own words, and lack of intentional instruction or training on paraphrasing. In a seminal study, undergraduate students were asked to identify properly paraphrased or plagiarized paragraphs. Most of the time students were able to identify correct paraphrasing; however, almost half of students characterized plagiarized text as properly paraphrased (Roig 1997). This study shows the difficulty students may have in understanding and recognizing paraphrasing and plagiarism and suggests the need to properly train novice writers to use information to support their ideas without copying (Roig 2015).

Steps can be taken by students to avoid plagiarism and develop as scientific writers. One method is intentional practice with note taking. When taking notes on text and ideas, include references and original sources to keep track of where information originated. Indicate within the notes text that came directly from the source (i.e., quotes) and which is your paraphrasing. A second method is to integrate information from multiple sources. Reviewing several sources of information allows you to develop the knowledge needed to properly paraphrase and apply the synthesized ideas to the specific writing project. The Academic Integrity Handbook for Students from Massachusetts Institute of Technology has additional guidelines to help promote appropriate paraphrasing to avoid plagiarism. They include:

- Altering paragraph order and individual sentence structure
- Switching between active and passive voice
- Replacing words and terminology to fit your writing style.

The last suggestion may be difficult for science writing as synonyms may not exist or be accurate for the terminology or jargon needed in your writing. Therefore, a string of specific terminology that cannot be avoided is acceptable and not necessarily seen as plagiarism in science writing. A far more thorough set of writing guidelines to avoid plagiarism can be found at the US Department of Health and Human Services Office of Research Integrity, authored by Miguel Roig (2015).

To put some of the above guidelines into practice, the following is an excerpt from a paper published by Heide Ford, a noted cancer researcher (Ford et al. 1998). The full reference is listed at the end of this chapter.

> To determine whether HSIX1 plays a role in regulating the cell cycle, the MCF7 mammary carcinoma cell line was transfected with SIXFL, a construct that allows for constitutive expression of the full-length wild-type HSIX1 cDNA, or with the parent vector expressing the chloramphenicol acetyl transferase gene (CAT) as a control. MCF7 cells were chosen because they are mammary carcinoma cells with an endogenous HSIX1 level far lower than that in 21PT cells (data not shown). Stable transfectants were selected by using cloning cylinders and examined for HSIX1 expression via Northern blot analysis (Fig. 2A). For all subsequent analysis, three stable clones expressing HSIX1 (HSIXA1, A8, and A13) and two control transfectants (CATB1 and CATB3) were examined.

An appropriate paraphrased version could be:

Ford et al. (1998) wished to assess the function of HSIX in breast cancer cell proliferation by creating stable cell clones overexpressing the gene. Experiments were conducted using three MCF7 cell lines stably overexpressing HSIX and two vector control cell lines containing chloramphenicol acetyltransferase.

An improper paraphrase of the original text would be:

(From Ford et al. 1998): To show whether human SIX regulates the cell cycle, MCF7 mammary carcinoma cells were transfected with a full-length version of HSIX to constitutively express the protein. As a control, cells were also transfected with the parent vector expressing the chloramphenicol acetyltransferase gene. MCF7 cells were chosen as these mammary carcinoma cells have lower endogenous expression of HSIX. Stable transfectants were selected and confirmed by Northern blot. Three HSIX experimental and two vector control cell lines were established and used for analysis.

The first paraphrasing option condenses the original text while maintaining the main points. Sentence structure and wording have been significantly changed, which required some content knowledge of the procedures involved such as making stable cell clones and overexpressing the target protein. In the improper version, some words were changed or omitted but the overall paragraph and sentence structure is intact. Despite citing the original source, the second option is bad practice for note taking and certainly shouldn't be included in a writing project unless it was appropriately indented and placed between quotation marks.

Solicit writing feedback from mentors, instructors, or other experts to provide guidance on acceptable citing practices and appropriate paraphrasing. Review and use self-reflection to intentionally avoid plagiarism. Develop the necessary knowledge required for your writing project and use the critical thinking methods within this book to reflect on your writing.

Finally, give yourself enough time for all the pre-writing and writing steps needed to complete your project.

The Best Way to Show Off: Data Presentation

One of the functions of scientific writing is to convey ideas and conclusions based on the analysis of data. Finding the most effective way to present your data can be a challenge and requires the same level of research and planning as writing. When presenting data, the author must decide what they wish to convey and how it might best be accomplished. What is the best format for data visualization? Is it a bar graph, scatterplot, or a multi-panel figure? Fortunately, resources are available to provide insight into the best approach for data visualization (Midway 2020; Nature Communications 2015). Researcher Stephen Midway has proposed ten principles that provide a strong framework to begin thinking about and designing how best to present data, which have been listed in table 14.1. Midway provides additional detail for each principle, and we will highlight some important points in the table.

For additional background information on the ten principles, see Midway's (2020) review as well an excellent series on data visualization from Nature Communications: *Visual Strategies for Biological Data, the Collected Points of View* (2015).

Conclusion

Science writing can be an overwhelming task. However, with guidance and intentional practice, writing can become a very effective method of communication. Thorough background research is needed to help develop useful outlines and avoid writer's block. "Write, review, and revise" should be a repeated process until a final draft of the writing assignment is complete.

Table 14.1

Ten Principles of Effective Data Visualization

1. Diagram first.	Sketch an outline of images and figures to critically assess how they should be displayed.
2. Use the right software.	What software do you need to best fit your needs? Photoshop? RStudio?
3. Use an effective geometry and show data.	Select the best format to display your data (bar graph? multi-panel figure?)
4. Colors always mean something.	Use color intentionally to enhance and not distract.
5. Include uncertainty.	Include statistics such as error bars or confidence intervals.
6. Use panels, when possible (small multiples).	If your data sets are complementary, place them into a multi-panel figure.
7. Data and models are different things.	Use raw data points whenever possible and distinguish from data models such as "best fit" lines.
8. Use simple visuals and detailed captions.	Figures and captions should stand alone from manuscript text.
9. Consider an infographic.	Simple labeling can increase the power of a graphic.
10. Get an opinion.	Seek out feedback on presentation design.

Source: S. R. Midway, "Principles of Effective Data Visualization," *Patterns* 1 (2020).

Intentional practice of putting new information into your own words will help avoid plagiarism and promote proper paraphrasing. For novice writers in science and medicine, this can be challenging. The continual development of artificial intelligence can be an attractive alternative but may still lead to plagiarism. Seeking out help and getting feedback is key. Finally, data visualization is every bit as important as the structure of your writing. Clear and effective figures can help relay information and strengthen your scientific arguments. Follow best practices for data visualization to complement science writing.

Summary

- Writing is an important form of communication in science and medicine.
- Do not underestimate the time it takes for a writing project.
- Write, review, revise, repeat.
- Practice paraphrasing for learning and to avoid plagiarism.
- Be cautious with the use of LLMs for writing and do not pass it off as your own work.
- Use published tips for data presentation best practices.

Exercises

1. Review a published article with several figures. Determine which data visualization principles are met in each figure. Suggest improvements that could be made to the figures to enhance their presentation.

2. Using the same article, practice paraphrasing using paragraphs in the methods or results sections.

3. Discuss in a peer group or with mentors how artificial intelligence can be appropriately used for writing and how to avoid the ethical pitfalls it can cause.

Works Cited

Council of Writing Program Administrators. 2019. "Defining and Avoiding Plagiarism: The WPA Statement on Best Practices." https://wpacouncil.org/aws/CWPA/pt/sd/news_article/272555/_PARENT/layout_details/false.

Ford, H. L., E. N. Kabingu, E. A. Bump, G. L. Mutter, and A. B. Pardee. 1998. "Abrogation of the G2 Cell Cycle Checkpoint Associated with Overexpression of HSIX1: A Possible Mechanism of Breast Carcinogenesis." *Proceedings of the National Academy of Sciences of the United States of America (PNAS)* 95: 12608–12613.

Knisely, K. 2005. *A Student Handbook for Writing in Biology.* Sinauer Associates.

Midway, S. R. 2020. "Principles of Effective Data Visualization." *Patterns Prejudice* 1: 100141.

Nature Communications. 2015. *Visual Strategies for Biological Data, the Collected Points of View (2010–2015).* https://www.ami.org/professional-resources/expert-methods.

Nelms, G. 2012. *Plagiarism Overview: What Research on Plagiarism Tells Us.* https://www.wright.edu/sites/www.wright.edu/files/page/attachments/PlagiarismOverview.pdf.

Quitadamo, I. J., and M. J. Kurtz. 2007. "Learning to Improve: Using Writing to Increase Critical Thinking Performance in General Education Biology." *CBE–Life Sciences Education* 6: 140–154.

Roig, M. 1997. "Can Undergraduate Students Determine Whether Text Has Been Plagiarized?" *Psychological Record* 47: 113–122.

Roig, M. 2015. *Avoiding Plagiarism, Self-Plagiarism, and Other Questionable Writing Practices: A Guide to Ethical Writing.* The Office of Research Integrity, US Department of Health & Human Services.

Schimel, J. 2012. *Writing Science: How to Write Papers That Get Cited and Proposals That Get Funded.* Oxford University Press.

15 A Blueprint for Learning: How to Prepare for Class, Bloom's Taxonomy, and Learning Objectives

> The most useful piece of learning for the uses of life is to unlearn what is untrue.
> —Antithenes

You may have noticed that textbook chapters and lectures often begin by stating "learning objectives." These objectives have a defined structure and purpose that might not be initially obvious; nonetheless, being able to read learning objectives and understand their purpose and goals will help you to understand which information is most important and how to study.

Defining Cognition and Metacognition

Before we discuss learning objectives and frameworks for their design, we need to define *cognitive* and *metacognitive thinking*. These terms are important in order to understand the frameworks used to write learning objectives and come into play when discussing study strategies in chapter 17 and information triage in chapter 18. Cognitive thinking consists of mental processes we use every day such as remembering the streets needed to drive home. For students, cognitive thinking would include processes such as predicting, summarizing, and comparing, all of which are especially important for studying and learning new material. On the other hand, metacognitive thinking, in the simplest of terms, can be defined as thinking about thinking. A metacognitive process is one in which you are aware of your own thoughts. Tasks such as planning, reflective practice, and revising require metacognition. These are tasks that ask you to consider your previous or current thought processes and what changes might need to be made to meet a specific goal or objective. A great deal has been written in education and psychology fields about cognitive and metacognitive thinking. However, for our purpose we can simply focus on how these thinking processes relate to learning objectives and how they should be approached by students.

What Are Learning Objectives and What Are They For?

Graduate and medical education programs include didactic learning sessions, such as lectures, problem-based learning sessions, or interactive laboratories. Educational expectations of these didactic classes are commonly expressed through learning objectives, which can then be connected to assessments.

Educators need to align the material to be taught with the material the students are expected to learn and the material that is assessed. Learning objectives help align these goals. Therefore, learning objectives are useful tools for both the instructor and the student.

Learning objectives are not to be confused with learning goals. Goals can be thought of as larger accomplishments for a program or course, while objectives are more specific and detailed tasks that can be measured, such as through multiple choice or written answer questions.

As an example, let's use a genetics course. A course goal might be to understand genetic inheritance and how it can lead to disease. This goal is general and simply asks to relate genetic inheritance to disease. The goal is phrased so broadly that no specific measure could assess whether it was achieved. However, a learning objective within the course might be to "compare and contrast inheritance patterns of autosomal dominant and recessive traits." This learning objective requires specific knowledge of the characteristics underlying selected patterns of genetic inheritance. It is specific and measurable, where you can imagine a test question asking about the distinction between an autosomal dominant and recessive family pedigree or about identification of characteristics related to either mode of inheritance.

This highlights a useful feature of understanding learning objectives for students: It allows you to predict which topics will be placed on examinations, and how the exam questions might be phrased.

Another advantage of understanding learning objectives for students in medical school or other professional graduate programs is it can promote critical thinking by placing responsibility for learning on the student. Well-composed learning objectives will increase comprehension by directing students toward a knowledge base and providing guidance on cognitive and metacognitive tasks needed to meet the objective. The "level" at which the learning objective is written (discussed in more detail below) will determine whether a student needs to memorize a series of facts, must apply previous knowledge to a clinical scenario, or is required to design an experiment based on scientific literature. The specific tasks targeted by a learning objective will also inform the type of preparation needed before a didactic learning session and what kind of study strategies might be most useful afterward. Despite the direction a learning objective can provide, there is still a great deal of freedom for a student to reflect and choose appropriate cognitive and meta-cognitive learning strategies (keep this in mind for chapters 17 and 18).

Learning Objective Frameworks

Multiple frameworks and formats are used for learning objectives. Three common formats for stating objectives found in education are categorized as either "I can" or descriptive, "Students should be able to," statements. The "I can" statement is typically found in K-12 education but can be useful for professional, higher education programs. They can serve as a checklist of sorts where a student can go through multiple learning tasks. Examples would be "I can identify the twelve sets of cranial nerves" or "I can classify amino acids based on side groups." However, common learning objective formats in medical education are descriptive, or "Students should be able to," statements. Both are similar in content with verbs utilized to express the level of expected knowledge. The specific genetics learning objective used above is an example of a descriptive statement but can be easily modified as a "Students should be able to" statement (students should be able to compare and contrast inheritance patterns of autosomal dominant and recessive traits). Along with the style in which the learning objective is written, specific frameworks are used to relay the thinking process needed to meet the objective. Two learning objective frameworks commonly found in medical education will be highlighted here: Bloom's taxonomy (Bloom 1956) and Quellmalz's framework for higher-order thinking (Quellmalz 1985). Several others have been created and are reviewed elsewhere (Moseley 2005).

Bloom's taxonomy was originally published in 1956 as a tool for educators to classify cognitive behaviors needed for different levels of learning. It consisted of six domains, which were *knowledge, comprehension, application, analysis, synthesis,* and *evaluation.* Although Bloom's taxonomy is often presented with a "pyramid" structure and hierarchy, with certain domains "higher" than others, this pyramid was not developed by Bloom and was never Bloom's intent. The taxonomy was intended as a means of allowing educators to communicate about educational goals by using shared definitions and to generate ideas for new education goals. For example, if one teacher's goal is for a student to "really understand" material, another's is for a student to "internalize knowledge about" material, and a third's is for a student to "comprehend" material, all three educators may have the same goal, but not shared language to express it. The analogy to a biological taxonomy is unfortunate because it causes the illusion that it was developed based on the natural relationships between educational goals. It cannot inform us of the relationship between the goals; it can only provide a common framework to discuss them.

Put another way, although certain tasks are often presented as "lower" in Bloom's taxonomy, such as rote memorization being lower than "synthesis," no cognitive science was done to determine if one task was simpler or less cognitively demanding than another. In reality, for any task we undertake in education, we are most likely exercising skills from multiple "domains."

Indeed, certain tasks, like recall of knowledge, which are frequently considered to be at the bottom of Bloom's pyramid, are cognitively complex and intertwined with other cognitive tasks. Cognitive scientist Daniel Willingham writes, "Data from the last 30 years lead [*sic*] to a conclusion that is not scientifically challengeable: thinking well requires knowing facts, and that's true not simply because you need something to think *about*. The very processes that teachers care about most—critical thinking processes like reasoning and problem-solving—are intimately intertwined with factual knowledge that is in long-term memory" (Willingham 2021). As mentioned in the chapter on cognitive biases (see chapter 3), many tasks we think of as "higher order" are in fact tasks of recall, such as a chess master who outperforms a novice only by having memories of a larger number of games to draw from. Placing recall at the bottom of the pyramid means that often educators will avoid testing recall to avoid objectives that are seen as "too simple."

Based on usage and educator feedback, the taxonomy was revised in 2001 (Anderson and Krathwohl 2001). In the revised version, the original taxonomy was split from one category into two, with the first category encompassing different types of knowledge structure and the second category including different domains of cognitive processes (table 15.1). The revision allowed for the separation of knowledge content from cognitive and metacognitive thinking processes. It also introduced verbs for the different cognitive processes, allowing educators to better connect and assess instruction and learning.

The revised Bloom's taxonomy and associated cognitive process verbs are an important tool for students. The verbs utilized in a learning objective point toward one or more cognitive processes needed to meet the instructor's learning expectations. Does a student just need to memorize a series of facts or will that student need to use those facts and apply them to a new scenario? The categories and verbs available through Bloom's

Table 15.1
Revised Bloom's Categories

Knowledge	Cognitive process with example verbs
Factual	*Remember* • Label, define, recall, identify
Conceptual	*Understand* • Explain, discuss, paraphrase, illustrate
Procedural	*Apply* • Utilize, devise, include, prepare
Metacognitive	*Analyze* • Classify, compare, contrast, distinguish
	Evaluate • Assess, critique, predict, justify
	Create • Produce, revise, plan, modify

classification function as a guide for the student. The use of Bloom's taxonomy as a reference for a particular lecture or lab can aid in determining the necessary breadth and depth of knowledge.

A second learning framework is the Quellmalz framework for higher-order thinking (Quellmalz 1985), which builds upon aspects of Bloom's classifications. The intended focus of this framework is on problem-solving with processes separated by "lower-order" or "higher-order" thinking (table 15.2). In this framework, the lower-order thought process involves recall, which aligns with remember and understand domains from Bloom's taxonomy. A student must possess previous knowledge for recall and thus is not necessarily creating new knowledge. The higher-order categories in the Quellmalz framework are seen not in a hierarchical progression of difficulty but as equal tasks necessary for analysis, evaluation, and creation. For

Table 15.2
Quellmalz's Higher-Order Thinking

Process	Category	Related bloom's category
Cognitive	Recall	Remember, understand
	Analysis	Analyze
	Comparison	Analyze
	Inference	Apply, create
	Evaluation	Evaluate
Metacognitive	Planning	Create
	Monitoring	Analyze
	Reviewing/revising	Create

these categories of higher-order thinking, a student must take previous knowledge and reconstruct it to fit a new question or problem. By emphasizing higher-order thinking, the Quellmalz framework pulls together philosophical and psychological definitions to promote critical thinking and problem-solving (Moseley 2005).

Both Bloom's and Quellmalz's frameworks are useful tools for instructors and students to clearly express learning expectations and the thought processes needed to achieve them. Although both frameworks present some thinking tasks as having greater difficulty than others, the relative importance of recall to evaluation can be debated and may be context-dependent. Perhaps a more appropriate approach from a student perspective to learning objective frameworks is to look at them through a heuristic lens. Use the classifications of cognitive and metacognitive processes as guides to take a degree of control and encourage self-directed learning. In a practical sense, that means look at the learning objectives to figure out what you are expected to study.

Learning Objectives: Before and After Class

Learning objectives can and should be utilized by the student both before and after a lecture or laboratory session. Before a didactic learning session, it is highly recommended that students prepare by reviewing learning objectives and other provided materials. This does not need to be an intensive, time-consuming process but a brief preview to become familiar with learning expectations. The preview will also introduce the cognitive strategies that will be needed to meet the objectives. The cognitive tasks come into play during and after the lecture or learning session. Going back to our hypothetical genetics lecture and learning objective, the verbs "compare" and "contrast" point toward cognitive tasks needed to promote the proper level of learning. Comparing and contrasting autosomal inheritance patterns can be accomplished by building upon previous knowledge (i.e., recall Mendel, pea plants, chromosomes) and putting the information into new formats, such as a table (i.e., analysis, comparison). By doing this, multiple cognitive thought processes are being used that are outlined in tables 15.1 and 15.2. Such strategies may seem like common sense to most students. However, for rigorous academic programs such as medical school, the quantity of information and the high level of academic expectations force many students into "survival mode." To get the most out of learning objectives and what they entail requires consistent and intentional effort by a student to maintain a high degree of success.

Summary

- Learning objectives are useful tools for both instructors and students to understand learning expectations.
- Learning objectives can be written in different formats and adhere to discrete frameworks.

- Regardless of the framework, learning objectives point toward common categories of cognitive learning strategies.
- Objectives are a key component to preview before a class and utilize afterward to ensure learning expectations are met.

Exercises

1. Revise the following objectives to promote higher-order thinking using Bloom's taxonomy (apply, analyze, evaluate, create):
   a. Know the function of mitosis and meiosis
   b. Understand the difference between the pathology of thrombotic infarct and embolic (hemorrhagic) infarct
   c. Define cytokines
2. Review a lecture from a previous course you have taken. If learning objectives are available, classify them using Bloom's taxonomy and Quellmalz's higher-order thinking framework. If no objectives were provided, compose appropriate objectives using one or both frameworks.

Works Cited

Anderson, L. W., and D. R. Krathwohl. 2001. *A Taxonomy for Learning, Teaching, and Assessing: A Revision of Bloom's Taxonomy of Educational Objectives*. Longman.

Bloom, B. S. 1956. *Taxonomy of Educational Objectives: The Classification of Educational Goals*. Longmans, Green.

Moseley, D. 2005. *Frameworks for Thinking: A Handbook for Teaching and Learning*. Cambridge University Press.

Quellmalz, E. S. 1985. "Needed: Better Methods for Testing Higher-Order Skills." *Educational Leadership* 43: 29–35.

Willingham, D. T. 2021. *Why Don't Students Like School?: A Cognitive Scientist Answers Questions About How the Mind Works and What It Means for the Classroom*. John Wiley & Sons.

> The test of a good teacher is not how many questions he can ask
> his pupils that they can readily answer, but how many questions
> he inspires them to ask him which he finds it hard to answer.
> —Alice Wellington Rollins

In higher education one method for the assessment of learning
is through testing, commonly from the use of single-best-answer
multiple choice questions (MCQs). This is especially true for pre-
professional or health education programs where licensing
exams are often in the MCQ format. Since many pre-professional
education programs utilize this format for high-stakes exams, it
is important for students to be familiar with the format and take
advantage of its use to promote learning. Students in graduate
programs or those developing their teaching skills will also benefit, since the MCQ format and rules given below are useful in
those domains as well.

Testing to Promote Learning and the Use of the MCQ

Formal testing in education has had a controversial past where
it was thought to inhibit learning, induce anxiety, and promote

superficial knowledge gains (Roediger and Karpicke 2006). However, experimentation has shown that testing can have a strong positive effect on student learning, referred to as the *testing effect* (Kuo and Hirshman 1996). Learning is promoted by spaced retrieval of information and feedback from correct or incorrect answers. The impact of the testing effect can be greater than simple restudy of material alone (Rowland 2014; Wheeler et al. 2003). A common method for testing is the use of MCQs, which have been shown to have similar validity and assessment outcomes as short-answer or other open-ended question formats (Palmer and Devitt 2007; Pham et al. 2018; Schuwirth and Van Der Vleuten 2004). MCQ exams can also test "higher-order" thinking skills and not just first-order rote memorization, making them effective and efficient methods of learning assessment (Palmer and Devitt 2007).

Focus has shifted in higher education from instructor-generated MCQs to engaging students in the practice of question writing. Positive learning growth can be seen when students use self-testing to evaluate whether they are meeting learning expectations (Dochy et al. 1999). Student-generated MCQs can promote learning and can be just as useful as instructor-generated questions (Bottomley and Denny 2011; Grainger et al. 2018; McLeod and Snell 1996). However, proper instruction and training for question writing must be provided to students for the process to be effective (Gillespie 1991; Papinczak et al. 2012). Additionally, the largest impact of student-generated question writing has been observed on lower-performing learners (André and Anderson 1978; Jobs et al. 2013). Student-generated questions can identify misconceptions, allowing for early intervention, and can promote "higher-order" thinking through composing complex, case-based context questions (Schuwirth and Van Der Vleuten 2004).

The Anatomy of a Case-Based Context MCQ

Context questions require test takers to use information within the question, frequently case-based, to solve a problem or make a decision. The case-based context MCQ commonly encountered in medical education has a standard format that includes a clinical case or vignette, a question lead-in, and answer options that include the correct answer and usually four distractors/wrong answers (figure 16.1). MCQs also include associated learning objectives and answer rationales. The learning objective helps focus the content to be assessed and determines the complexity of the exam item based on the desired cognitive task of the objective (such as Bloom's category). The question rationale explains why the correct answer option is correct and why the remaining distractors are incorrect. The rationale is useful to provide feedback to the test taker, which is an important aspect of learning through testing. Exam items can be categorized into first-, second-, or third-order questions based on

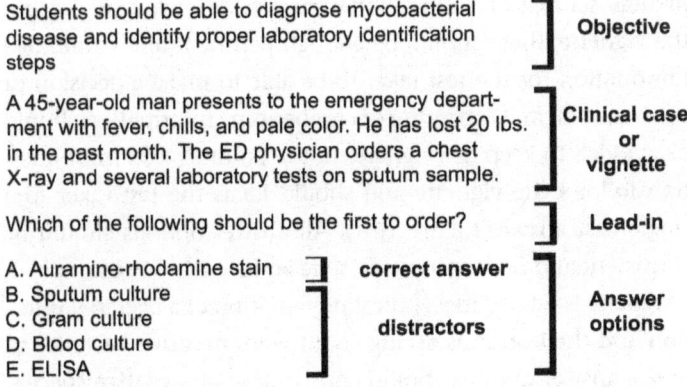

Figure 16.1
Anatomy of a quality MCQ exam item.

the level of knowledge needed to successfully answer the item. First-order questions are basic recall questions that would fulfill the "remember" category of Bloom's taxonomy. Second- and third-order questions require two or three steps from the question information to derive the correct answer, respectively. The higher-order structured questions tend to target more complex cognitive tasks found in Bloom's categories.

Detailed outlines of how to construct a case-based single best answer MCQ have been published elsewhere, and we encourage the reader to utilize those resources (Billings et al. 2020). Here we will briefly outline the major points to writing a case-based MCQ and common writing flaws to avoid.

What Goes into a Case-Based MCQ?

The first step to writing any test question is to determine the learning objective and the cognitive category to be tested. Once an appropriate learning objective has been identified, a short vignette can be written based on a clinical scenario or experimental setup (for foundational science knowledge). Within the vignette, there should be enough pertinent and contextual information for the test taker to be able to make a decision or solve a problem. Distracting or cumbersome information should be avoided to keep the vignette to the point. A lead-in or question follows the vignette and should focus the test taker to a single best answer (figure 16.1). All answer options should be consistent and homogenous in style and length. For example, if a vignette is related to a clinical presentation of a bacterial infection and the lead-in is asking about gram-negative bacteria, all of the answer options should consist of gram-negative species. When a case-based MCQ is written properly it should meet the "cover-the-options" rule. This is where a test taker can derive the correct answer based solely on the vignette and lead-in without the presence of answer options. Utilizing the cover-the-options rule helps test knowledge and reduce cueing, where the test-wise examinee can simply identify the correct answer among the selection of incorrect distractors (Schuwirth et al. 1996).

Common MCQ Writing Flaws

Writing a case-based MCQ can seem like an overwhelming task. However, by using the logical structure outlined above and avoiding common mistakes, question writing can be a useful learning tool. Below is a list, which is by no means exhaustive, of common writing flaws to avoid. These flaws can prevent the exam item from accurately assessing knowledge or can tip a test-wise examinee.

Grammatical Errors

Grammatical errors, especially in the lead-in question, can help eliminate answer options and provide clues to the correct answer without having any knowledge of the topic. Best practice is to word the question such that grammatical cues do not impact the answer options.

Example 16.1 A 45-year-old man presents to the emergency department with fever, chills, and pale color. He has lost 20 lbs. in the past month. The ED physician orders a chest X-ray and several laboratory tests. The first laboratory test ordered was an:

i. Auramine-rhodamine stain

ii. Sputum culture

iii. Gram stain

iv. Blood culture

v. ELISA

In English the indefinite article "an" is used when the following word begins with a vowel, and "a" is used when the following word begins with a consonant. So a student paying attention can eliminate answers B, C, and D, because only answers A and E begin with vowels. Thus, they are being tested in part on their knowledge of the English language and not the material being taught.

Absolute or Vague Terms

A common mistake when writing a vignette is to use vague terms such as "commonly" or "usually." These terms will not be interpreted the same by everyone. Instead, be specific with frequency, such as "twice daily" or "once per month." Another mistake with terminology is within the answer options. Distractors that include absolute terms such as "all" or "never" should be avoided as they are less likely to be correct.

Example 16.2 A 45-year-old man presents to the emergency department with fever, chills, and pale color. He has lost 20 lbs. in the past month. The ED physician orders a chest X-ray and several laboratory tests. Which of the following is the common recommendation to treat this infection?

i.   A regimen of rifampin

ii. A combination of isoniazid, rifampin, pyrazinamide, and ethambutol

iii. Isoniazid and rifampin only

iv. Pyrazinamide and ethambutol for 2 months

v.   All of the above

Overly Detailed Answer Options and More Detailed Correct Answers

In many instances, the test writer will put their focus on making sure the correct answer option is accurate and spend less time on the remaining distractors. When this occurs, the correct answer will be longer and more detailed. The wise test taker will detect this and guess the correct answer. A similar error is when all of the answer options are long and overly complicated. The proper format for the case-based MCQ is for the vignette and lead-in to carry the weight. The answer options should be short and concise so that they measure knowledge and not reading ability. Shorter answer options also fit with the "cover-the-options" rule stated before. Answer options should be short, concise, and consistent, with similar structure and length.

Example 16.3   A 45-year-old man presents to the emergency department with fever, chills, and pale color. He has lost 20 lbs. in the past month. The ED physician orders a chest X-ray and several laboratory tests. Which of the following is the next step in management of this infection?

 i. Combination of extended antimicrobial treatment
 ii. Confirm laboratory identification using a nucleic acid assay and culture
 iii. Isolation of the patient
 iv. Quarantine of the patient
 v. PPD skin test followed by sputum culture

Convergence and Word Repetition (Clanging)

Convergence is a writing flaw where the correct answer has the most in common with the other distractors. Therefore, the correct answer can be chosen by determining what terms converge together. Clanging, or word repetition, is where words within the vignette are repeated in the correct answer option. This is an obvious clue for a test taker, and the question will not be useful to assess learning. Reviewing and editing answer options will allow for this writing flaw to be avoided.

Example 16.4 A 45-year-old man presents to the emergency department with fever, chills, and pale color. He has lost 20 lbs. in the past month. The ED physician orders a chest X-ray and several laboratory tests. Which of the following is the common recommendation to treat this infection?

i. A combination of isoniazid, rifampin, pyrazinamide, and ethambutol

ii. Rifampin and pyrazinamide

iii. Isoniazid and rifampin

iv. Pyrazinamide and ethambutol

v. Pyrazinamide and ioniazidIn this example, there is convergence. The patient most likely has tuberculosis. Each answer contains some tuberculosis medications, so the student can guess that (i) is the correct answer.

Pseudo-Case and Teaching in the Vignette

In context case-based MCQs, the vignette and lead-in should work together and both be sufficient and necessary to select the correct answer. The lead-in needs to ask a question that requires information within the vignette to be answered correctly. A question writer does not want to spend the effort of writing a strong vignette just to write the following lead-in: "Which of the following amino acids has a negative side chain?" Any information in the vignette is irrelevant as the lead-in stands alone.

A second flaw commonly found in questions is teaching in the vignette. This is when the test writer includes new information on a topic and uses this information to help guide the test taker.

Example 16.5 A 45-year-old man presents to the emergency department with fever, chills, and pale color. He has lost 20 lbs. in the past month. The ED physician orders a chest X-ray and several laboratory tests. The physician suspects a bacterial infection that doesn't stain with the Gram stain. The first step in the identification of *Mycobacteria* is:

i. Auramine-rhodamine stain

ii. Sputum culture

iii. Gram stain

iv. Blood culture

v. ELISA

Most high-stakes MCQ exams also avoid "none/all of the above" answer options, negative lead-ins such as "which of the following except," true/false items, and overly complicated vignettes. Inclusion of these types of question characteristics can reduce the reliability and validity of the assessment.

Understanding what goes into a case-based context MCQ will help remove anxiety associated with this type of assessment format. As a student, you can then focus on knowledge acquisition and application. Additionally, trying to write quality MCQs can be an excellent learning exercise to help solidify knowledge. The ability to create a well-written case-based question that is challenging requires critical thinking and deep content knowledge, both of which are critical to the growth of graduate and professional students.

Conclusion

Assessment or testing is an important step in the learning process. It can help solidify knowledge as well as highlight areas of focus for further study. Multiple choice questions are commonly encountered in professional education programs and can have similar high-value learning outcomes as other forms of examination. Learning to recognize and write quality MCQs is a valuable skill for both test taking and learning. Understanding how a MCQ is constructed along with recognizing common writing flaws can help in composing high-quality exam items that can be used for effective self-study.

Summary

- Assessments can promote learning through critical analysis and feedback.
- The testing effect from exams can have a more profound effect on learning than restudy.
- Writing effective and well-constructed exam items can be a method of study.
- There are specific methods for quality MCQ writing, especially for medical education.
- There are specific writing flaws that should be avoided so the assessment is testing knowledge and not helping test-wise learners guess the correct answer.

Exercises

1. Identify the flaws and revise the example items below.

    Example 1: A 23-year-old female presents to her family physician with fever, cough, and fatigue that started few days ago. An

ELISA test was positive for influenza. Which antibody is most effective at preventing an influenza infection?

a. IgA

b. IgD

c. IgE

d. IgG

e. IgM

Example 2: A 54-year-old male presents with an overly active inflammatory response as confirmed by a laboratory assay. The physician suspects that he may have an underlying inflammatory disease such as ulcerative colitis. Which of the following cytokines would likely be elevated in this condition?

a. TNF-α

b. IFN-α and IL-10

c. IL-10 and TGF-β

d. TGF-β and IFN-α

e. TGF-β and TNF-α

2. Using the information in this chapter as a guide, write a case-based context MCQ on a favorite topic such as biochemistry or physiology. Share with other students for peer feedback.

Works Cited

André, M., and T. H. Anderson. 1978. "The Development and Evaluation of a Self-Questioning Study Technique." *Reading Research Quarterly* 14: 605.

Billings, M. S., K. DeRuchie, K. Hussie, et al. 2020. "NBME Item-Writing Guide: Constructing Written Test Questions for the Health Sciences." Preprint.

Bottomley, S., and P. Denny. 2011. "A Participatory Learning Approach to Biochemistry Using Student Authored and Evaluated Multiple-Choice Questions." *Biochemistry and Molecular Biology Education* 39: 352–361.

Dochy, F., M. Segers, and D. Sluijsmans. 1999. "The Use of Self-, Peer and Co-Assessment in Higher Education: A Review." *Studies in Higher Education* 24: 331–350.

Gillespie, C. S. 1991. "Questions About Student-Generated Questions." *Journal of Reading* 34.

Grainger, R., W. Dai, E. Osborne, and D. Kenwright. 2018. "Medical Students Create Multiple-Choice Questions for Learning in Pathology Education: A Pilot Study." *BMC Medical Education* 18: 201.

Jobs, A., C. Twesten, A. Göbel, et al. 2013. "Question-Writing as a Learning Tool for Students—Outcomes from Curricular Exams." *BMC Medical Education* 13: 89.

Kuo, T.-M., and E. Hirshman. 1996. "Investigations of the Testing Effect." *American Journal of Psychology* 109: 451–464.

McLeod, P. J., and L. Snell. 1996. "Student-Generated MCQs." *Medical Teacher* 18: 23–25.

Palmer, E. J., and P. G. Devitt. 2007. "Assessment of Higher Order Cognitive Skills in Undergraduate Education: Modified Essay or Multiple Choice Questions?" *BMC Medical Education* 7: 49.

Papinczak, T., R. Peterson, A. S. Babri, et al. 2012. "Using Student-Generated Questions for Student-Centred Assessment." *Assessment & Evaluation in Higher Education* 37: 439–452.

Pham, H., M. Trigg, S. Wu, et al. 2018. "Choosing Medical Assessments: Does the Multiple-Choice Question Make the Grade?" *Education for Health* 31: 65–71.

Roediger, H. L., and J. D. Karpicke. 2006. "The Power of Testing Memory: Basic Research and Implications for Educational Practice." *Perspectives on Psychological Science* 1: 181–210.

Rowland, C. A. 2014. "The Effect of Testing Versus Restudy on Retention: A Meta-Analytic Review of the Testing Effect." *Psychological Bulletin* 140: 1432–1463.

Schuwirth, L. W. T., and C. P. M. van Der Vleuten. 2004. "Different Written Assessment Methods: What Can Be Said About Their Strengths and Weaknesses?" *Medical Education* 38: 974–979.

Schuwirth, L. W. T., C. P. M. van der Vleuten, and H. H. L. M. Donkers. 1996. "A Closer Look at Cueing Effects in Multiple-Choice Questions." *Medical Education* 30: 44–49.

Wheeler, M., M. Ewers, and J. Buonanno. 2003. "Different Rates of Forgetting Following Study Versus Test Trials." *Memory* 11: 571–580.

17   Academic Toolbox: Study Strategies, Time Management, and Group Dynamics

> Tell me and I forget; teach me and I may remember; involve me and I learn.
> —Xunzi

Advanced education such as professional or graduate school is one of the most challenging undertakings any learner can do. The amount of information in the biomedical field has been increasing exponentially, and how you approach this challenge and become successful is critical. For many learners, doing what they have already done in undergraduate school simply will not suffice. Open-mindedness and changes are needed to excel. Change is never easy but is necessary. Learners must swiftly adopt evidence-based learning strategies, time-management skills, and group studying. In this chapter we are going to explore all these methods and discuss key elements that have been shown to be effective.

The Myth of Learning Styles

The purpose of education is to promote learning, which requires a great deal of effort on design and implementation.

One method in education is to match instruction to a student's individual learning style. This may be through oral, visual, or kinesthetic means. Surveys have shown that people, including educators, believe that learning occurs best through individual learning styles and that they are static, meaning they don't change (Nancekivell et al. 2020). You may have stated yourself that you learn best by reading or by working with your hands. Unfortunately, an individual possessing a dominant learning style is a prevalent myth in education, one that is not supported by evidence (Pashler et al. 2008).

Research has shown that students with different self-identified learning styles don't perform differently (Husmann and Loughlin 2019; Pashler et al. 2008). Student self-reported learning styles do not match their actual study habits, nor do they have any influence on exam performance (Husmann and Loughlin 2019). Perpetuating the myth of learning styles may actually do more harm than good by keeping student focus away from other highly effective methods (Willingham et al. 2015). Learners have been shown to be poor at recognizing evidence-based learning strategies that do not align with their self-identified learning style (McCabe 2011).

So how does one break free of the learning styles myth and move forward to enhance learning? One way is through explicit instruction. Educate yourself on evidence-based learning strategies through either specific courses or through researching the topic yourself. Intentional instruction on learning has been shown to increase student ability to identify evidence-based learning strategies (McCabe 2011). Another way to move past learning myths is to promote what is called "desirable difficulty" (Bjork and Bjork 2011). This concept refers to keeping your studying "appropriately difficult" to promote long-term learning. Encouraging difficulty in studying helps fend off the idea of learning styles that tend to be easier for students to implement. Desirable difficulty, however, only works if the learner has sufficient background knowledge and skill. Without the right foundation, the learning task becomes inappropriately difficult and can lead to undesirable outcomes. An inappropriate difficulty would be to ask a group of students with no background in cardiac physiology to review a lecture on the topic and identify the most important medically relevant themes and concepts. Without prior knowledge on the topic, it will be incredibly difficult to accomplish such a task and may cause frustration. The action is also unlikely to foster long-term retention. Fortunately, there are evidence-based methods to achieve desirable difficulty during directed self-learning.

Evidence-Based Study Strategies

The process of learning is an active one (Freeman et al. 2014). Many learners engage passively with information and deploy ineffective, repetitive strategies to gain knowledge, such as using flashcards, rereading text, or reviewing presentation slides. Learners instead need to acquire information using *active learning* strategies such as self-assessment, spaced-retrieval, interleaving, and varied practice. This allows the learner to retain the information, develop a deeper understanding, and be able to recall the information over a longer period of time. Pursuing active learning and desirable difficulty will keep you from taking shortcuts such as using premade tables or charts instead of creating them.

- *Self-assessment:* The use of *practice questions* for self-assessment was discussed in detail in chapter 16. The "testing effect" can be a powerful learning tool, and several studies have shown that testing can improve learning and retention with the percentage of retention after one week varying from 21 to 44 percent (Bjork 1988; Karpicke and Roediger 2008; Roediger and Karpicke 2006; Runquist 1983). Practice questions can help align what has been learned with the educational objectives of a lecture or course.

- *Spaced retrieval:* Studying is more effective when done intermittently, a method called *spaced retrieval* or spaced practice. Block study (cramming) is inferior in both short- and long-term retention (Bjork and Bjork 2011). To implement spaced-retrieval, study a topic over several days using active learning approaches. For example, if you are taking a biochemistry course you would study every day for several days or weeks before the exam. During each day you would study for an hour, take a break, self-assess, then list what you know and what you don't know. You could then follow this up with another hour of active study, a short break, self-assessment, and another knowledge summary. The next day you can then start with the previous day's summary, followed by more cycles of study.

- *Interleaving:* Interleaving is when you study different topics together as opposed to studying materials in blocks. Instead of just taking a biochemistry course, let's add a histology course as well. For most students, the natural thing would be to use one day to study biochemistry and the next day to just study histology. Interleaving would be studying both biochemistry and histology during the same day. Using our example of spaced retrieval, you would study biochemistry for an hour, then histology for an hour, followed by self-assessment of both courses. Essentially, you are going back and forth between different topics. Although this is not intuitive, studies have shown that interleaving performs better than block study on exams and provides superior long-term retention (Kornell and Bjork 2008). Interleaving is hypothesized to work by forcing your brain to compare and contrast topics as you study them together. This means you are working within the "analysis" category of Bloom's taxonomy (chapter 15) and engaging "higher" cognitive functions (Bjork and Bjork 2011).

- *Varied practice:* Studying that becomes routine and predictable can promote superficial or false learning (Bjork and Bjork 2011). Let's say you study a set of vocabulary words every night for a week at the same time and location by reading the definitions repeatedly. You then assessed your "learning" by using flash cards. You most likely would be able to memorize your vocabulary words, and your self-assessment would give you the impression you learned them. However, if you were asked to write a fictional story using all of the vocabulary words two weeks after you studied them, you most likely would struggle to remember them, let alone use them in the correct context. This is where varied practice can be used to enhance learning. The idea here would be to change your study approach. Even studying the vocabulary words in a different room will increase retention (Smith et al. 1978). Instead of predictable routines, change where and how you study. To put the concept of varied practice into context for graduate and medical students, utilize active learning to create concept maps, information tables, drawings, or flow charts. Change where you study and with whom. This may seem unnerving since routine can provide comfort; however, varied practice fits with desirable difficulty.

- *Explanation/elaboration:* This strategy, like the others discussed, can be done individually or as a group. Elaboration occurs when you connect new information with previous knowledge. The process of taking new information, integrating it with what you already know, and then putting it into your own words helps solidify your knowledge. An example would be paraphrasing, as discussed in chapter 14. Additionally, visual or verbal strategies can be used such as creating a diagram or explaining a concept. Elaboration promotes deeper comprehension as it requires metacognitive categories of analysis, evaluation, and creation of a revised Bloom's taxonomy (Anderson and Krathwohl 2001).

Group Dynamics

One way to implement many of the above strategies such as varied practice and elaboration is through group study. Many careers, especially in science and medicine, require the ability to effectively collaborate and work as a team. Therefore, professional and medical education programs utilize small group active learning curricula in the form of problem-, case-, or team-based exercises. Working in a group is an effective tool to enhance individual learning performance. Finding or creating a group of classmates for group study can be an excellent method to improve deep learning. Before you begin to set up study groups, it is important to be aware of some of the best practices to make sure your group has a positive and effective experience.

Five stages have been identified as groups learn to work together. These stages, based on work by Tuckman and Jensen (1977), are forming, storming, norming, performing, and adjourning. Table 17.1 shows the general functions with each stage in group development. A functional group will go through all of these stages, which may overlap with each other. Awareness and understanding of the stages of group development can help promote positive dynamics.

Table 17.1

Group Stage	Function
Forming	Meet and get a feel for group members
Storming	Some conflict from personality differences Conflict resolution
Norming	Create ground rules and reasonable expectations
Performing	Do the work to meet group's goals
Adjourning	Finish the job and move on

Several characteristics that strongly impact learning from group work are communication, engagement, openness, support, and type of dominant behavior (Dionne et al. 2020).

1. First, *communication* helps establish relationships and reduce uncertainty. This characteristic was found to be most crucial for the forming and performing stages of group development.

2. *Engagement* is how willing members are to participate and hold each other accountable, which can increase group performance. Engagement was found to be especially important during the forming stage.

3. *Openness* is critical to allow for respectful interactions and to have all group members' thoughts and opinions heard. It is mutually necessary with strong engagement and is fostered by quality communication. Openness will also lend itself to the next characteristic, which is support. Positive and effective groups are dependent on each member of the group.

4. *Supportive* interactions and communication create a trusting and safe learning environment. Without it, members can feel rejected, feelings can be hurt, and the dynamics of the group can become negative.

5. A negative *dominant* voice is enough to create a negative experience. Conflict may arise during the forming and storming stages, which inhibits positive communication, openness, and support, thus crushing any hope of effective performance. It is important to be aware that your sole actions can impact the learning of everyone in the group. However, not all dominant behavior is bad. There is strong support that positive-dominant group members can promote all of the above characteristics and enhance group performance (Dionne et al. 2020). This is accomplished by allowing all group members to have their opinions heard and respected. Having a dominant group member or members is not itself a bad thing as long as they are open and promote positive interactions.

These factors are also affected by the size of a group. A large group may allow some members to avoid participating without notice, but in a small group the personalities of individual members are more likely to dominate.

Time Management

No one can effectively deal with the quantity of information provided to them in their advanced education without developing a strategy for managing time. This is borne out by research suggesting that time management is a better predictor of medical student success than aptitude tests such as the MCAT. Time management is also a strong predictor of perceived success in graduate school. The following is a list of some techniques for time management.

1. *Develop a schedule.* One way to start is to create a semester-long schedule, then break that down to weekly schedules,

then to daily schedules. Remember that your schedule needs to have some breaks, and time to eat. Practice checking your schedule every morning and throughout the day until it is integrated into your daily routine as a habit.

2. *Use checklists.* To avoid forgetting tasks that need to be accomplished and falling behind, begin adding tasks to a checklist (or to-do app) as you become aware that they need to be accomplished. Set deadlines for yourself and hold yourself accountable.

3. *Manage time.* As you work, ask yourself, When is this due? How long will it take me to complete and can I start it now? How long have I been working on this and do I need to work on the next thing? Do I need to go now so that I can get to my next commitment? "Chunk" your tasks. For tasks that take more time than can be dedicated to a single sitting, you will need to plan out what "chunks" of the tasks can be accomplished in each sitting. For example, if you are writing a book about critical thinking, you might choose one day to write 1,000 words, another day to reread and edit two chapters, and another day to search out references.

4. *Analyze your own patterns.* If you are consistently falling behind or failing to accomplish your tasks, examine how your time is being used, what is causing you to lose time, and how it can be altered to accomplish the desired tasks. If last week you were unable to study for microbiology at all because you reviewed your physiology lectures three times, then repeating your pattern again the next week will result in the same result. A change is required to produce a different and potentially a more positive outcome.

5. *Set timers.* Every smartphone comes with a timer app. Use it to time your tasks and to remind you to move to the next

one. One popular study method is the pomodoro method, which uses timers (some of which are shaped like tomatoes) to interleave study time with breaks and topic changes to keep the mind engaged.

6. *Sleep.* Sleep disturbances prior to an exam correlate with decreased exam performance among medical students. Sleep timing may be more important for medical exam performance than the actual sleep length or quality; for example, disruptions to the circadian clock (such as by pulling an "all-nighter") can disrupt exam performance. For this reason it is important to develop and stick to a sleep schedule. Avoiding sleep prior to exams to "cram" is counterproductive because any additional study time is counteracted by the deleterious effects of the sleep disruption.

7. *Seek feedback.* If your time management is not working for you, show your schedule to an academic advisor, coach, or friend and ask for suggestions. The feedback process may improve your schedule and your ability to stick to it.

Conclusion

It is critical that students in professional programs reconsider their prior study strategies, group studying, and time management as they embark on this new adventure.

Summary

- Self-educate about evidence-based learning strategies.
- Find the appropriate level of difficulty to challenge your brain during studying.
- Find or create a study group.

- In group study, find the appropriate group size to make sure it is big enough for diversity of thought but not too big to manage.
- Create and promote a positive dynamic in group studying.
- Use tested methods for time management.
- Don't forget to sleep and eat on a regular basis to keep your body and brain fresh.

Exercises

1. In a spreadsheet, make a weekly schedule for yourself that will help you manage your time effectively. Does it account for accomplishing all of your tasks? Does it allow adequate breaks and time for sleep?
2. Reflect on your previous study habits. Which evidence-based approaches aligned best?

Works Cited

Anderson, L. W., and D. R. Krathwohl. 2001. *A Taxonomy for Learning, Teaching, and Assessing: A Revision of Bloom's Taxonomy of Educational Objectives*. Longman.

Bjork, R. A. 1988. "Retrieval Practice and the Maintenance of Knowledge." In *Practical Aspects of Memory: Current Research and Issues, vol. 1: Memory in Everyday Life*, edited by M. M. Gruneberg, P. E. Morris, and R. N. Sykes, 396–401. John Wiley & Sons.

Bjork, E. L., and R. A. Bjork. 2011. "Making Things Hard on Yourself, but in a Good Way: Creating Desirable Difficulties to Enhance Learning." In *Psychology and the Real World: Essays Illustrating Fundamental Contributions to Society*, edited by M. A. Gernsbacher, R. W. Pew, L. M. Hough, and J. R. Pomerantz, 56–64. Worth.

Dionne Merlin, M., S. Lavoie, and F. Gallagher. 2020. "Elements of Group Dynamics That Influence Learning in Small Groups in Undergraduate Students: A Scoping Review." *Nurse Education Today* 87: 104362.

Freeman, S., S. L. Eddy, M. McDonough, et al. 2014. "Active Learning Increases Student Performance in Science, Engineering, and Mathematics." *Proceedings of the National Academy of Sciences* 111: 8410–8415.

Husmann, P. R., and V. D. O'Loughlin. 2019. "Another Nail in the Coffin for Learning Styles? Disparities Among Undergraduate Anatomy Students' Study Strategies, Class Performance, and Reported VARK Learning Styles." *Anatomical Sciences Education* 12: 6–19.

Karpicke, J. D., and H. L. Roediger. 2008. "The Critical Importance of Retrieval for Learning." *Science* 319: 966–968.

Kornell, N., and R. A. Bjork. 2008. "Learning Concepts and Categories: Is Spacing the 'Enemy of Induction'?" *Psychological Science* 19: 585–592.

McCabe, J. 2011. "Metacognitive Awareness of Learning Strategies in Undergraduates." *Memory & Cognition* 39: 462–476.

Nancekivell, S. E., P. Shah, and S. A. Gelman. 2020. "Maybe They're Born with It, or Maybe It's Experience: Toward a Deeper Understanding of the Learning Style Myth." *Journal of Educational Psychology* 112: 221–235.

Pashler, H., M. McDaniel, D. Rohrer, and R. Bjork. 2008. "Learning Styles: Concepts and Evidence." *Psychological Science in the Public Interest* 9: 105–119.

Roediger, H. L., and J. D. Karpicke. 2006. "The Power of Testing Memory: Basic Research and Implications for Educational Practice." *Perspectives on Psychological Science* 1: 181–210.

Runquist, W. N. 1983. "Some Effects of Remembering on Forgetting." *Memory & Cognition* 11: 641–650.

Smith, S. M., A. Glenberg, and R. A. Bjork. 1978. "Environmental Context and Human Memory." *Memory & Cognition* 6: 342–353.

Tuckman, B. W., and M. A. C. Jensen. 1977. "Stages of Small-Group Development Revisited." *Group & Organization Studies* 2: 419–427.

Willingham, D. T., E. M. Hughes, and D. G. Dobolyi. 2015. "The Scientific Status of Learning Styles Theories." *Teaching of Psychology* 42: 266–271.

18 The Art of Information Juggling

The art of being wise is the art of knowing what to overlook.
—William James

The volume of biomedical information has been increasing exponentially for the last several decades, and this trend is continuing (Densen 2011). To a student who needs to learn significant volumes of material in the short period of an advanced education, such as the first two years of medical or graduate school, it can feel like attempting to drink from a firehose. As a result, developing the skills to be able to process, prioritize, and sort all that information is vital for success. We're calling it "information triage," from the French *trier*, meaning to sort, and by analogy to the way a battlefield physician must make rapid decisions about which patients require care most urgently. Physicians constantly utilize this skill in prioritizing patient treatment based on the severity of their condition, history of present illness, medical and family history, laboratory tests, and physical exam findings. The skill of information triage is not inherent and needs to be learned through intentional practice. For students, developing this skill and eventually mastering it is essential to academic success.

Cognitive Load

Cognitive load is the term for the workload placed on working memory resources by attempting to hold more pieces of information than the *working memory* has "slots" for (Cowan 2001).

In cognitive load theory, when we "learn" something it moves from working memory to long-term memory, and as our understanding deepens we may develop "schemata," which are more complicated representations of topics.

In chapter 3 we used the example of learning the name of every single tree we encounter anew versus developing a *schema* that allows us to recognize something as tree-like due to its shape, color, bark texture, leaves, and other qualities. In this sense a schema is like a compressed form of complex information that can be held in working memory in lieu of every element of the schema.

When we are presented with too much information at once and we don't have adequate schemata, our working memory can become overloaded, and learning becomes less effective.

In the cognitive load model, learning occurs when the *element interactivities* of complex tasks and ideas have been collapsed into schemata that can be held in working memory. Elements are units of "things" that need to be learned or understood. Low element interactivity concepts are things that can be learned with minimal outside reference, such as learning the symbols of chemicals on the periodic table. You don't need to hold the symbol for boron in working memory while learning the symbol for neon. A high element interactivity concept is one that requires many interacting elements to be held in working memory. For example, an algebraic equation like $(a+b)/c=d$, where we need to solve for a, would require someone learning algebra to hold each symbol in working memory [a, b, +, (,), /, c, =, d] as well as the rules for order of operations and for solving equations. A more advanced learner who has developed appropriate schemata can solve the problem by the use of heuristics and schemata such as by intuitively "knowing" that the first step is to rearrange to $a+b=dc$.

The natural complexity of the material is called the *intrinsic* element interactivity, and additional unnecessary complexity created by the way that material is presented is termed *extraneous* element interactivity. Well-designed curricula, lesson plans, and activities minimize the extraneous element of interactivity so that students aren't presented with more information than they can handle at once. Poor curricular design will introduce unnecessary difficulty.

For example, imagine two students who are trying to solve the above equation. Student 1 did all of their homework in a previous math class and has internalized the "PEMDAS" concept for order of operation (parentheses, exponents, multiplication and division, and addition and subtraction). Student 2 was distracted in the previous year by difficulties at home, and did not develop this schema. The instructor saves time by not reviewing this concept before assigning this problem. Student 1 is able to solve the problem with much greater ease than student 2, despite being assigned the same problem.

Developing necessary schemata before moving on to more complex material makes learning the new material much easier. Reading a book is a complex and difficult task if you must sound out every word, look up unknown words, and consult someone about sentence structure. However, once you have learned to become a strong reader, the task becomes easy because schemata exist for reading vocabulary and grammar. This is why reading skills are emphasized before other skills that depend on the ability to read.

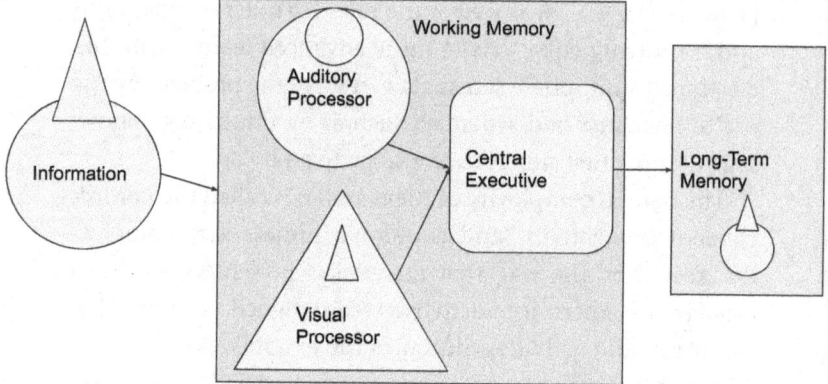

Figure 18.1
Cognitive load theory: Information is taken into the working memory through the senses, processed, and stored in long-term memory. These memories can later be retrieved to short-term memory; however, the short-term memory can only manage a limited number of items at a time. Managing cognitive load prevents the working memory from becoming overloaded.

This presents you with a challenge when presented with a lot of new information, some of which may cause extraneous cognitive load. Suppose you need to interpret the flow of ions in the nephron, but you haven't developed a strong schema for membrane transport. You will have to do the extra work of looking up how membrane transport works and entering it into short-term memory, and you will have less working memory slots for other parts of the material.

If you previously studied membrane transport and developed enough understanding to have a useful schema, then learning the new material will be much easier, and may even appear "intuitive" to another student observing who never developed the membrane transport schema.

Your challenge in your education is that you're likely to be presented with instructional design that introduces extraneous cognitive load beyond the natural complexity of the material, or you may be expected to start your program with schemata in place that you didn't learn during your undergraduate education.

Your goal isn't to memorize every individual fact; it's to move complicated tasks and understandings from conscious to unconscious processing by building schemata through practice over time.

Educators can minimize cognitive load by practicing *scaffolding*, which is the practice of developing the necessary schemata prior to introducing concepts that rely on those schemata.

As a student, your challenge is to deal with both the intrinsic and extraneous cognitive load that accompanies the material you are presented with. The remainder of this chapter addresses exercises to develop skills in tackling the curriculum and doing your own scaffolding (Kirschner et al. 2009).

Strategies for Information Processing

An important first step when starting a new course or program is to identify the learning expectations. Using learning expectations will allow you to determine what new information is of greater importance and how to integrate it with what you already know. The following are strategies to help identify and classify the vast amount of information that gets presented within graduate and medical education.

1. *Identify learning objectives:* Before starting your studying, it's important to identify the learning objectives (chapter 15). Coupling the information to the learning objectives allows you to focus on the necessary information that needs to

be acquired and prevents you from spending study time in ineffective ways (Anderson and Krathwohl 2001).

2. *Scan the material:* You should start by quickly scanning the material to get an overview of the topics covered. This will aid you in acquiring a general idea of the content and identifying any portions that may require further attention. It is necessary that you identify headings, subheadings, and key concepts that are central to the understanding of the information.

3. *Use active reading strategies:* As you browse the information and identify the most important concepts, use active reading strategies to engage with the material. Using evidence-based techniques can improve the efficiency of the study process. Certain techniques like highlighting text are popular, but are not supported as particularly effective. Effective strategies may include constructing tables or charts, developing concept maps, summarizing, or taking brief notes. When note taking, remember to properly paraphrase information (chapter 16). Connect new information with what you already know. These strategies will help you stay focused and promote effective retention.

4. *Organize the information:* It is important that the information is organized into categories or themes. This allows integration of the information and can show how the many different pieces coalesce together. This is where creating an outline, a mind map, or concept map can be an effective way for organization as well as visualization of the connections between different concepts.

5. *Assess and relearn:* The use of questions to assess your understanding of the material is vital for retention as has been shown by multiple studies (see chapter 16). This breaks up

the monotony of simply reviewing the information repeatedly and also allows the learner to identify knowledge gaps and ways to close those gaps.

6. *Identify missing pieces to scaffold:* If you are struggling to understand material, it may be that you have not yet developed the necessary schemata to understand the material. In order to scaffold the material for yourself, you will need to understand which tasks and processes you are expected to understand and which you are not yet proficient in, and focus study time on those topics.

Assembling Information

Next, we will turn our attention to an active process that will help you discover how to manage and triage information in a practical setting. You may do this either individually or with your study mates. Assemble the information that you are trying to learn, such as a recorded lecture, a chapter in a book, a supplemental text, or any other learning information. Next, ask yourself the following questions.

1. What is the first thing you would/should do? *In answering this, consider: How does this relate or fit into other information you have already learned? It is important to anchor information to existing knowledge to enhance retention.*

2. What are the external resources that you would use? Why? How do you use a textbook? *In addressing this question, make sure to use reliable sources and ones that align with the learning objectives provided to you. Sometimes students venture into resources that do not align with the required objectives and material and waste precious time learning information that is not relevant to the task at hand.*

3. What are the key points of this lecture? *Here is where you would create headings, subheadings, and key concepts and go beyond the objectives to sub-objectives.*

4. How would you determine what is important in this lecture? *This is at the heart of the learning objectives and an important skill to acquire to be successful. Knowing which information is the most important in a large amount of knowledge is vital. You would also benefit from other people's perspectives in your study group.*

5. How would you integrate this information with the existing information you know? *Any new learned information must be anchored to existing information and build on what you already know. This might be harder in the beginning, but as more knowledge is gained and a schema is built, the easier it will become. Being able to find practical application of the information is also important for retention and usefulness.*

6. How would you summarize this information if you need to review it? *Organizing the information as we have discussed is important for learning, reviewing, and retaining. This is achieved by building tables and diagrams, flowcharts, concept maps, or mind maps. There are several ways of organizing information, as stated, but different subjects may require different approaches. For example, a table would work well for listing microorganisms, the diseases they cause, lab tests results, and other characteristics, but a concept map would be better to organize the interconnectedness of the different lymphocytes, the cytokines they produce, and their function.*

7. How did this lecture help you meet my learning goals and what could be improved to advance your learning? *This is a good place to reflect on what has been learned and how you will next use the information.*

8. How would you design your study plan for this material? *There are multiple methods to tackle this, and there isn't a right or wrong answer. This may also be field-dependent; what might work for anatomy may not work for biochemistry. For example, flash cards may work well for remembering anatomical structures and their function, while drawing pathways might work better for the subject of biochemistry.*

9. How would you determine that you have learned the material? *This is a very important step. Learners need to assess how well they have learned the material. The evidence shows that both writing and using questions work best for assessing knowledge, especially if spaced between learning the information.*

10. How do you plan on retaining this information? *Many of the suggestions above will help with this. Think of the many things that have been discussed and the ones you came up with to address this.*

Conclusion

Triaging information while studying involves identifying learning objectives, scanning the material, using active learning strategies, organizing the information, prioritizing it based on relevance, and assessing and relearning the most important information. By applying these strategies and using the questions above to guide you, you can effectively manage large amounts of information, which will allow you to focus your studying and put you on the path of success.

Summary

Triaging information is a challenging task that learners face as the amount of information is exponentially increasing. This can be addressed by deploying effective learning strategies, such as:

1. Identifying learning objectives

2. Scanning the material

3. Using active reading strategies

4. Organizing the information

5. Assessing and relearning.

Exercises

Incorporating what you have read about study and learning strategies, apply the following questions to a lecture or course. Use them as a guide to determine the best approaches to studying and learning the material.

1. What is the first thing you would or should do?

2. What are the external resources that you would use? Why? How do you use a textbook?

3. What are the key points of the lecture?

4. What is the relevance of this topic to medicine or science?

5. How would you integrate this information with the existing information you know?

6. How would you summarize this information if you need to review it?

7. How did this lecture help you meet your learning goals, and what could be improved to advance your learning?

8. How would you design your study plan for this material?

9. How would you determine/assess that you have learned the material?

10. How do you plan on retaining this information?

Works Cited

Anderson, L. W., and D. R. Krathwohl. 2001. *A Taxonomy for Learning, Teaching, and Assessing: A Revision of Bloom's Taxonomy of Educational Objectives*. Longman.

Cowan, N. 2001. "The Magical Number 4 in Short-Term Memory: A Reconsideration of Mental Storage Capacity." *Behavioral and Brain Sciences* 24: 87–114.

Densen, P. 2011. "Challenges and Opportunities Facing Medical Education." *Transactions of the American Clinical and Climatological Association* 122: 48–58.

Kirschner, P. A., F. Kirschner, and F. Paas, 2009. "Cognitive Load Theory." https://core.ac.uk/download/pdf/55535268.pdf.

19 Grow Through Feedback: Evaluative Judgment

The only real mistake is the one from which we learn nothing.
—Henry Ford

Imagine a refrigerator that tried to stay cold without measuring temperature, an ant that attempted to navigate to food without smelling, or a student who tried to learn a skill without receiving feedback. No process that requires maintaining or improving any quality can be conducted without assessing it. Feedback is the mechanism by which we understand what we have learned and what more we have *to* learn.

Feedback is a critical component of learning. It helps promote self-reflective and self-regulatory processes to induce change. Students are exposed to various feedback resources, such as quizzes, exams, peers, and instructors that can be relayed through written, verbal, or automated sources. There is a disconnect between students and teachers about what feedback is meant to do, how it should be implemented, and what makes it effective (Tai et al. 2018). The goal of this chapter is to help realign your thinking about feedback so it can become an integral part of learning in graduate or professional programs. We will discuss the instructor's role in the feedback process and how the instructor–student interaction can promote self- and peer-evaluation.

Feedback Literacy: Knowledge for the Student

Giving or receiving feedback can be a difficult experience with potentially unintended consequences, such as confusion or hurt feelings. With preparation and intentional design, feedback can be productive and rewarding.

For feedback to be useful to students, they must first realize what feedback is, what is needed to make use of it, and acknowledge their role in the feedback process (Carless and Boud 2018). One purpose of constructive feedback is to help "learners increase their skill set while working to achieve the expected outcomes. It raises their awareness about their performance and directs their future actions" (Ramani and Krackov 2012). Ideally, a student should be able to take in feedback, reflect on it, and then use it to modify behavior to enhance future performance. This idealized model of feedback is not always realized, as students and many instructors are not literate in feedback. There is a disconnect between student and instructor perceptions on feedback (Tai et al. 2018). Although both students and instructors viewed feedback as a method to promote improvement, the focus of each group was different. Students felt that individually tailored and actionable feedback was a priority, while instructors put a greater emphasis on format and design of the feedback experience (Dawson et al. 2019). Interestingly, in a study by Dawson and colleagues that evaluated student and instructor perceptions on feedback, there was no mention of critical aspects of feedback such as using feedback for evaluation, the use of exemplars, and the effective use of peer feedback (Dawson et al. 2019). Students and instructors agree on the purpose of feedback, but their focus is on different elements, and both groups are missing important aspects of the feedback process. Feedback literacy aims to realign student and instructor thinking and promote effective strategies.

A framework to promote feedback literacy has been developed for students that includes four elements: *feedback appreciation*, *making judgments*, *affect management*, and *taking action* (Carless and Boud 2018). Feedback literacy is also promoted through specific training modalities, such as peer review and analysis of exemplars (Camarata and Slieman 2020).

Feedback Appreciation

Feedback is not a passive process with unidirectional flow from instructor to student. Effective feedback should be an active and engaging process for students. Students should seek out feedback and not wait passively for it to occur. Think about the design and delivery elements of feedback opportunities created by instructors. Does the design of the feedback process engage you as a student? Does it produce the type of actionable feedback necessary to promote change? Feedback can come in many forms, which may be specifically designed, such as through written, verbal, recorded, or automated (such as with quizzes and exams) platforms. Self-evaluation is a key aspect of using feedback to measure current performance against what should be clear expectations. Any gap that is identified can be the target for improvement on future learning assessments.

Making Judgments

Feedback is used to foster self-reflection and judgment. Both characteristics are needed to convert feedback to an action plan that will improve performance. Development of self-judgment can be enhanced by student engagement, self-assessment, participating in peer review, and analysis of exemplars (Boud et al. 2013; Carless and Boud 2018). The goal is for you to become accurate at evaluating yourself and your performance. Does it measure up to the academic or professional standard? Be honest and critical of your own work, as well as the work of others.

Affect Management

Receiving feedback, especially if it is perceived as negative, can elicit an emotional response that can disrupt its value and effectiveness. It is important to manage emotional responses and not view critiques as personal (Carless and Boud 2018). A safe and trusting environment between the student and instructor needs to be established (Krackov 2013; Ramani and Krackov 2012). From the instructor's side, it is important that feedback be based on direct observation and delivered using specific and respectful language. From the student side, manage reactions to critiques, stay engaged and promote dialogue, and reflect on the information provided following the feedback session. Even negative feedback can be appropriately received if both the student and instructor work to create a productive relationship. The proper relationship can allow for the exchange of ideas and engage the student as an active participant in the feedback process.

Taking Action

For feedback to have effect, students need to be receptive to information, analyze it, and modify behavior for future performance. This requires specific skills to properly understand and utilize feedback, such as self-motivation. Opportunities with feedback that promote self-motivation can be ineffective if the instructional design limits the student's ability to show the feedback was effective. Feedback becomes actionable if it is targeted toward self-regulation and specific processes (Hattie and Timperley 2007). Generalized feedback on personal traits may be difficult to interpret, while task-specific comments may not translate to unrelated tasks. Therefore, it is important for students to be aware action is needed and to develop the self-motivation for continuous improvement.

Models of Feedback

Several feedback models have been developed, and we high-light three that are commonly used, especially in medical education. The first model that most students are probably familiar with is called the *feedback sandwich* (or *compliment sandwich*). This is where negative feedback is sandwiched in between positive statements and is typically part of a unidirectional flow of information from the instructor to the student. It is not an effective strategy, as the intended purpose can be lost due to the back-and-forth nature of how statements are delivered (Molloy et al. 2020). The sandwich model does not actively engage the recipient, as they are simply the receiver of feedback, and it does not fit current conceptual and experimental models of effective feedback.

Two feedback models that are more student-centered and appear to be more effective are the *educational alliance feedback model* and the *relationship, reactions, content, and coaching* (R2C2) model (Sargeant et al. 2018; Telio et al. 2015). The educational alliance model relies on the quality of the relationship between the student and instructor. The student works with the instructor (providing the feedback) and has an awareness of a mutual understanding of the feedback goal. There is reciprocal respect between the student and instructor, which allows for agreement with how the feedback goal can be achieved (Telio et al. 2015). The educational alliance model is based on the perception of credibility of the instructor by the student, and allows for a two-way flow of information. This results in mutually respectful interaction and clear delivery of critical feedback. The R2C2 model is based on a similar premise of student engagement and relationship with the person providing feedback (Sargeant et al. 2015). The four phases of the model include building a relationship between the student and instructor, exploring reactions to and perceptions of the provided feedback, ensuring the content of the feedback is understood, and coaching to promote change. Question prompts are provided for each step to guide the instructor in meeting the objectives of each phase in the process (Sargeant et al. 2015). Both the educational alliance and R2C2 feedback models are based on a mutually respectful relationship and have the student receiving the feedback as an active participant.

Closing the Loop

The most effective models of feedback place the student or individual receiving the feedback as a central and active figure in the process. The models are also meant to be reiterated, allowing for opportunities to put the feedback into action followed by reassessment. This is termed the *feedback loop*, where a student performs a task, the task is compared with a standard, feedback is provided to compare the student's task to the standard, the student then makes and carries out an action plan, and reassessment occurs, hopefully with improved outcomes. It is important to close the feedback loop as an instructor and provide opportunities for students to act on feedback. Without closing the feedback loop, the process is ineffective and credibility from the student perspective breaks down. The feedback loop can be put into more practical terms. For example, a graduate student submits a manuscript for publication, and the reviewers ask the student to revise the submission. The student can then act on those reviews and resubmit the manuscript. The reviewers could then accept the manuscript or provide additional feedback. A similar scenario can take place in medical education where a medical student performs a newly learned clinical skill. Feedback is provided on how well the student performed the skill, and the student is provided with another opportunity to show they understood and acted on the feedback. Closing the feedback loop is a critical step toward effective feedback practices.

Criteria for Effective Feedback

For feedback to be properly received and effective, several characteristics should be met. A safe and respectful environment needs to be created between the student and instructor. Goals and objectives should be jointly agreed on and clearly communicated. Clear and specific communication allows for the comparison of student self-assessment with the shared objectives. During feedback encounters, emotional responses should be managed by both the student and instructor. Instructors should use appropriate language and provide feedback on directly observable behavior. Following feedback discussions, it is important to ensure the receiver of the feedback understands and accepts the critiques. Once feedback has been accepted, a jointly created action plan can be created. Student self-motivation along with instructor coaching is key to following through on the action plan. A reiterative process needs to be established to close the feedback loop and assess how well the feedback has been implemented. Two final criteria for effective feedback include timeliness and reflection. Feedback for a task or process under evaluation needs to occur in an appropriate time frame so that the feedback is useful and can be implemented. An example would be getting feedback on a writing assignment the day before it needs to be turned in. There may not be sufficient time to understand the feedback, let alone include suggestions in a final draft for submission. Without timeliness, feedback has little usefulness. Finally, all participants in the feedback process should reflect on the experience. Were criteria for effective feedback met? How could you modify your role to improve the process next time? Self-reflection on one's role in feedback helps maintain integrity and usefulness of the process.

Evaluative Judgment

Evaluative judgment is a cognitive skill utilizing critical assessment, feedback, and development of mastery to promote lifelong learning. Evaluative judgment can be defined as "the capability to make decisions about the quality of work of self and others" (Tai et al. 2018). Students need to move beyond passively receiving information on quality and become active and independent learners, to go from directed self-learning to self-directed learning. This evolution requires the development of evaluative judgment where one can accurately identify the quality of work without relying on the judgment of others.

For students, *evaluative judgment* is needed to appropriately act on feedback. The student and instructor are engaged together to agree upon a common set of goals and objectives. Whether the student meets the objectives is determined by both evaluation from the instructor as well as the student's self-evaluation. If the student is not capable of accurate self-assessment, then the feedback becomes disconnected, and the process can break down.

Accurate self-evaluation also appears to correlate with academic performance. One study compared student self-assessments with those of their teachers over multiple exams, in multiple courses, over multiple semesters (Boud et al. 2013). Comparison of student self-assessment with the independent teacher assessment at the first exam in a course showed a significant difference. Students overestimated their performance on first exams in a single course. On subsequent exams, student and teacher assessments became comparable, suggesting students became better at self-evaluation over time. Analysis of first course exams over multiple courses continued to show student overestimation on performance, which again became comparable to teacher assessments on subsequent exams in each course. The study went further and divided students into groups of overestimators, underestimators, and accurate estimators of exam performance (Boud et al. 2013). The student self-assessment category was then correlated with actual academic performance throughout the study period. Overestimators tended to be poor achievers academically, while those who underestimated or accurately estimated their scores had higher academic achievement. Further, students who were more accurate with self-evaluation showed the greatest level of academic improvement over time. The students who consistently overestimated their performance were typically poor academic achievers (Boud et al. 2013).

The development of accurate self-evaluative judgment is critical to not only academic success but also professional success. Once you have begun your career stage of life, it is up to you to determine the quality of your work. There may be professional standards to guide you, but they may not be as explicit as in a formal classroom setting. Strategies are available to help develop evaluative judgment, some of which have been mentioned above (and discussed in Tai et al. 2018). One strategy is the analysis of exemplars. These are examples that can be provided by instructors to show different levels of quality. The ability to see examples of quality as well as comparing with others can be helpful in understanding standards. The use of rubrics can also help promote evaluative judgment and align with exemplars. Critical use by a student of a rubric of their work and of peers can provide a basis for quality of work. The use of rubrics can be applied to peer review, another strategy for developing proper self-judgment. The combination of exemplars, rubrics, and peer review can be implemented to increase uptake and quality of feedback (Camarata and Slieman 2020; Slieman and Camarata 2019). Intentional use of the above strategies can help develop strong judgment skills, which are critical for both learning and future professional success.

Conclusion

Feedback is a key to learning. However, for students to accept and act on feedback, some level of knowledge of the topic is required. A minimal level of feedback literacy can help promote effective strategies toward receiving and utilizing feedback. The strategies of effective feedback also aid in the development of self-evaluative judgment, a critical skill for academic and professional success.

Summary

- Feedback is critical for learning.
- Feedback literacy is important for students to accept and act on feedback.

- For students, feedback should be an engaging process where they are part of the conversation of mutually agreed-upon goals with the instructor.
- Specific criteria should be followed to ensure successful feedback encounters.
- Several feedback models exist, but the most useful are student-centered.
- Feedback, peer review, and analysis of exemplars lead to the development of evaluative judgment.

Exercises

1. Reflect on an experience where you received feedback from an instructor. What kind of feedback model was used? Can you identify any of the design elements that were implemented by the instructor? Were any of the criteria listed above followed? How effective did you find the experience?

2. Design a peer-review experience with a classmate. Include the necessary feedback criteria, along with a way to close the feedback loop. Following the feedback episode, reflect on the experience. How difficult did you find the process? Were you able to engage your classmate in the feedback experience? What would you do differently?

Works Cited

Boud, D., R. Lawson, and D. G. Thompson. 2013. "Does Student Engagement in Self-Assessment Calibrate Their Judgement over Time?" *Assessment & Evaluation in Higher Education* 38: 941–956.

Camarata, T., and T. A. Slieman. 2020. "Improving Student Feedback Quality: A Simple Model Using Peer Review and Feedback Rubrics." *Journal of Medical Education and Curricular Development* 7: 2382120520936604.

Carless, D., and D. Boud. 2018. "The Development of Student Feedback Literacy: Enabling Uptake of Feedback." *Assessment & Evaluation in Higher Education* 43: 1315–1325.

Dawson, P., M. Henderson, P. Mahoney, et al. 2019. "What Makes for Effective Feedback: Staff and Student Perspectives." *Assessment & Evaluation in Higher Education* 44: 25–36.

Hattie, J., and H. Timperley. 2007. "The Power of Feedback." *Review of Educational Research* 77: 81–112.

Krackov, S. K. 2013. "Giving Feedback." In *A Practical Guide for Medical Teachers*, edited by J. A. Dent and R. M. Harden, 323–332. Churchill Livingstone.

Molloy, E., R. Ajjawi, M. Bearman, et al. 2020. "Challenging Feedback Myths: Values, Learner Involvement and Promoting Effects Beyond the Immediate Task." *Medical Education* 54: 33–39.

Ramani, S., and S. K. Krackov. 2012. "Twelve Tips for Giving Feedback Effectively in the Clinical Environment." *Medical Teacher* 34: 787–791.

Sargeant, J., J. Lockyer, K. Mann, et al. 2015. "Facilitated Reflective Performance Feedback: Developing an Evidence- and Theory-Based Model That Builds Relationship, Explores Reactions and Content, and Coaches for Performance Change (R2C2)." *Academic Medicine* 90: 1698–1706.

Sargeant, J., J. M. Lockyer, K. Mann, et al. 2018. "The R2C2 Model in Residency Education: How Does It Foster Coaching and Promote Feedback Use?" *Academic Medicine* 93: 1055–1063.

Slieman, T. A., and T. Camarata. 2019. "Case-Based Group Learning Using Concept Maps to Achieve Multiple Educational Objectives and Behavioral Outcomes." *Journal of Medical Education and Curricular Development* 6: 2382120519872510.

Tai, J., R. Ajjawi, D. Boud, P. Dawson, and E. Panadero. 2018. "Developing Evaluative Judgement: Enabling Students to Make Decisions About the Quality of Work." *Higher Education* 76: 467–481.

Telio, S., R. Ajjawi, and G. Regehr. 2015. "The 'Educational Alliance' as a Framework for Reconceptualizing Feedback in Medical Education." *Academic Medicine* 90: 609–614.

Conclusion

We started this book with the assumption that you are driven by passion. You want to refine that passion into useful skills and knowledge through the process of learning. The most important takeaway from this book is that relying on intuition and learned mental shortcuts isn't always a bad thing, but those shortcuts need to be developed through an effortful process of reflective and critical thinking. The effortful process refines our thinking so that we can bypass biases, navigate knowledge, hatch hypotheses, elucidate ethics, gather instructional insights, calibrate cognitive capacity, and eventually develop new habits of mind.

Critical thinking can be understood as a kind of "system 2" thinking, which means slow, reflective, and deliberate thinking. In contrast, "system 1" is fast, intuitive, and relies on pre-encoded schemata and biases. Learning about critical thinking doesn't mean you will always practice system 2 thinking, but your effortful thinking will be more fruitful when you do.

The cognitive process of learning involves encoding items from working memory into long-term memory schemata. This repository of knowledge, experiences, and approaches can be used for better future system 1 thinking. Critical thinking allows you to evaluate, and potentially replace, your old schemata and biases with new schemata. This allows ways of thinking that are more useful and better aligned with rationality. You can identify and rectify misconceptions, mistaken beliefs, and biases. The result? Better decision-making and a better understanding of the world. Integrating critical thinking into your learning process can improve your mental models in the long run.

Index

Publisher contact:
The MIT Press
Massachusetts Institute of Technology
77 Massachusetts Avenue, Cambridge, MA 02139
mitpress.mit.edu

EU Authorised Representative:
Easy Access System Europe, Mustamäe tee 50,
10621 Tallinn, Estonia
gpsr.requests@easproject.com

Printed by Integrated Books International,
United States of America